Oak-Framed Buildings

Oak-Framed Buildings

Revised Edition

Rupert Newman

Published 2014 by

The Guild of Master Craftsman Publications Ltd,

166 High Street, Lewes, East Sussex BN7 1XU

ISBN: 978-1-86108-726-3

First published in 2005; reprinted 2007, 2010

ISBN: 978-1-86108-379-1

British Cataloguing in Publication Data. A catalogue record of this book is available from the British Library.

Publisher: Jonathan Bailey
Production Manager: Jim Bulley
Managing Editor: Gerrie Purcell
Senior Woodworking Project Editor: Stephen Haynes
Editor: James Beattie
Managing Art Editor: Gilda Pacitti
Design: Phil & Traci Morash at Fineline Studios

Typeface: Mahsuri Sans
Colour reproduction by GMC Reprographics
Printed in China

About the Author

Rupert Newman has been actively involved in carpentry since the age of 12. Having completed a degree in Naval Architecture he began working as a shipwright, but soon realized that building roofs and other large structures was his true passion. He is now the owner of Westwind Oak Buildings Ltd, a well-respected company that employs a number of highly skilled craftsmen and apprentices, building modern structures with green oak in the traditional way. From his base in the south-west of England, he undertakes building projects all over Britain and Europe.

Acknowledgements

I would like to thank my family, Lyndsay, Tom and Rowan, for all the support they have given me whilst writing this book. Without their help and encouragement it would have been an impossible task. Also thanks to all the carpenters I have worked with, especially Steve May, Oliver Campbell, Jon Lewis, Jude Green, Aron Marky, Kasper and Adrian Sharpe for their help with this book and for making such a great team. I am very grateful to Wilf Burton, architect and friend, for the supply of some superb drawings. I would like to acknowledge the input from Dan O'Neil for his research and copy reading, John and Jane Thompson for architectural and planning advice respectively, Frederick Madeline for his forestry information, Jon Loy for help with US zoning laws and Simon Ely for his glazing details. Thanks also to GMC Publications, especially James Beattie, Gerrie Purcell, Sara Harper and Anthony Bailey. Also to all the clients who so graciously allowed me to use photographs of their properties, I cannot thank them enough. Finally I owe a debt of gratitude to everybody at Westwind, to James Nicholson and Anne Tarren in the office, and all the boys and girls in the workshop.

Contents

Introduction

I have been very fortunate during my professional career to have been involved in the craft of traditional timber framing. When I started work as a young carpenter nearly 30 years ago, I never dreamt that I would end up dedicating most of my professional life to the practice of this ancient craft. Over the course of all those years, the techniques of oak framing have become better understood. As an industry, it has grown from a small, alternative form of construction, into a well-recognized building technique that employs large numbers of craftspeople. It is these people who have made it such an interesting and rewarding occupation for me, and it is to them that I owe the greatest debt. I have been fortunate to work with some fantastic carpenters, and a number of them have become like my family. Building an oak frame ultimately requires a team effort, and without the help and support that I receive from the members of my team I would be able to build nothing.

The idea of writing this book came from the need to answer some of the many questions that I get asked about my trade. These questions come not only from the clients I'm building frames for, but also from architects and other professionals involved in the design, the builders employed on the general construction and even the carpenters who work with me. Very little has been written, especially for the layperson, about building modern green-oak-framed buildings in the traditional style. Most of those books that are already in print focus on historical oak framing. This book concentrates on the unique and traditional British technique of oak-frame construction, a style of designing, building and erecting buildings which is rapidly gaining in popularity on both sides of the Atlantic.

Throughout the course of this book I have tried to shed some light on the process of designing and constructing an oak-framed house. It covers the properties of oak in some detail, as well as how frames are designed, made and raised, before examining the final stages of construction. *Oak-Framed Buildings* is not intended to be a step-by-step manual; rather its purpose is to provide readers with some knowledge of the techniques involved so they may better understand the overall process. Whether you are looking for ideas before embarking on a self-build project, are a member of the industry or working towards a job in timber framing, or perhaps just want to know more about this fascinating ancient tradition, I hope that this book provides interest and inspiration alike.

What is green-oak framing?

Before we delve into the details of joint construction or the characteristics of oak, let's start with the basics of what green-oak framing is. Well, it's a method of making a structural frame that will support both the roof and the floors of a building and on which the external envelope can be hung. The beams within the frame are connected using traditional joints and secured with wooden pegs. In North America this method would simply be known as 'timber framing' but unfortunately in Britain this generally means houses built with 4x2in (100x50mm) softwood, nailed together. To distinguish 'heavy' structural timber framing from 'light' softwood timber framing, 'oak' tends to be included somewhere in its description. So whether it's described as green-oak timber framing, traditional oak framing or just oak framing it all means the same thing. Confusingly though, oak is not the only timber used to make structural frames, even if it is by far the most common. I have concentrated on oak throughout this book but that is not to dismiss alternatives. Other timber, such as Douglas fir, can make equally impressive frames. Its properties are different to those of oak, so consequently it is framed in a different way. For instance, joint and section sizes tend to be bigger and curves are not generally used. Houses built using Douglas fir tend to have a different feel and style, more similar to American 'post-and-beam' framing than to British-style oak frames.

Throughout the book I refer to building frames for complete houses. In reality I am asked to build many other types of structure as well. The techniques described can equally be applied to building a conservatory, an extension, a bridge or even a swimming pool enclosure.

Left **The majestic sweep of an arch-braced truss highlights just one of the aspects of an oak-framed building that makes them so attractive**

Above right **While this book will hopefully provide you with plenty of interest and inspiration, erecting large oak frames is a skilled job and should only be undertaken by fully trained personnel**

Right **Oak frames can take many forms and do not always need to be complete buildings in their own right. This picture shows an oak-framed extension**

Oak framing, like most trades within the construction industry, can be dangerous if not done correctly. Building large structural frames is a science that takes years of training to become proficient in. Frame designs should always be passed by a qualified structural engineer before construction starts. Not only that, but the frames themselves should only be fabricated and erected by skilled and experienced carpenters. The usual route into the craft is through 'on the job' training, and apprenticeships are also offered by larger companies. The Carpenters' Fellowship has now set up an apprenticeship programme and a range of qualifications in Structural Post and Beam Carpentry, which is run through the Oak Frame Training Forum (www. carpentersfellowship.co.uk). I hope that these routes produce the next generation of carpenters to continue and further the craft. For more information on oak-framing courses and apprenticeships, see Useful Contacts on page 187.

Whilst writing this book I've been trying to describe (hopefully with some success) how to design and build oak frames. It has been a long journey for me, from my early days as a site carpenter working alone, to building massive oak frames with lots of other carpenters. During that journey I have learnt that there is always more than one way of doing things, and that should be borne in mind when reading this book. I have described how I would build a frame, how I would raise a frame, but not necessarily how other people might do it. If there are any errors or inconsistencies, they are all mine. I have depicted my style of building, which is very much individually designed and handmade. There are other ways of making frames – such as using mass-production techniques including CNC (Computer Numerically Controlled) machines – which I haven't touched on in this book as I'm not qualified to do so. I hope you find the following informative and that it goes some way towards answering your questions about the art of oak timber framing.

Previous page **Although this book concentrates on oak as the primary building material, other timbers, as this Douglas-fir frame demonstrates, make equally wonderful structures**

Right **An oak rafter is hoisted up a valley**

Far right **Oak-framed buildings do not have to conform to a rectangular plan. This dodecahedron extension shows that there are many design possibilities**

Chapter 1

A Brief History of Timber Framing

The history of timber buildings is incredibly long and varied; to cover it in depth would require several books. However, it is useful to know how some of the most important moments in its development have affected modern practice. Whilst the design of timber-framed structures has slowly evolved, many techniques used to construct them have remained constant. Many of these techniques are still in use today, and they define British-style oak framing.

Primitive Shelters

The first timber-framed shelters were made by primitive humans (*Homo erectus*) perhaps as long as half a million years ago, by tying sticks together and covering them with animal skins. These initial frames were probably used as temporary structures whilst out hunting, until the technology was further developed to create more substantial dwellings. The basic principle of construction was to make an 'A-frame' by lashing three pieces of wood together to form a rigid shape. Once a series of these frames had been made, they could then be raised in a row to make a simple house.

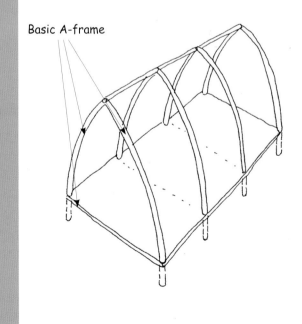

Basic A-frame

Right **An A-frame structure of this type was used in the creation of the earliest dwellings known to humankind**

Left **The Llandoger Trow in Bristol, UK is a classic example of a seventeenth-century timber-framed building**

Far left **A town house with jetties on two adjoining sides**

The next stage in development was to increase the height of the building by placing the roof-type structure on legs or posts. In order to do this accurately, the A-frames would have to be erected in position and joined together at their feet by a kind of timber ring beam or wallplate. Posts of the correct length to form the walls were sunk into the ground, and made firm by backfilling the earth. The roof structure was then reassembled onto the posts to form the building. This gave a skeleton on which the outer weather-stopping layer could be hung – in much the same way as a modern timber-frame building is made.

Iron-Age Structures

Post-hole buildings were common in Iron-Age construction and could be adapted to different shapes. The most common shape was the roundhouse in which archaeologists believe many Bronze- and Iron-Age people lived. Evidence of this has been the remains of post holes found in the ground and sometimes a circular trench left by the water dripping off the conical roof. Once the main timbers of a roundhouse had been erected, smaller coppiced material, such as hazel, would have been woven in between the posts and rafters, to form a more rigid structure. The walls would then have been covered with a mixture of daub, probably made of cow dung, straw and mud. This would have dried

to a very hard consistency, providing a sound walling material. Likewise, roofs would have been thatched with water reed or wheat straw, or alternatively, in some areas, a layer of turf.

Whole families lived together in roundhouses, and archaeological investigations have concluded that they would have been quite comfortable places to live, keeping warm in the winter and cool in the summer. A multitude of different carpentry techniques have been discovered on preserved timbers, such as mortise-and-tenon joints and scarf joints, along with the tools used to make them. The disadvantage of building posts into the ground was that it made them vulnerable to rot, but it was still a commonly used method of construction for peasant dwellings right up to the nineteenth century. In fact farmers still construct pole barns today by sinking telegraph poles into the ground.

Technological Development

When the Romans invaded Britain they brought with them many new construction techniques. Their jointing of timber was much more advanced than that of the native population and their technology included pegged mortise-and-tenons and more sophisticated triangulated roof trusses. However, when Rome's influence waned in the early fifth century this new technology seems to have been lost

and the during the subsequent Saxon and Viking rules construction reverted back to the post-hole technique, using mainly lap joints and notches. Mortise-and-tenon joints didn't really reappear until the reign of Henry II, which began in 1154. Henry II's reign marked the start of the Plantagenets' rule, which lasted until the death of Richard III in 1485 and held sway over a vast area from the Pyrenees, through Gascony to Normandy and over the Channel through England and to the borders of Scotland. During this period Europe became more stable, the climate improved and harvests were more plentiful. This led to an increase in wealth, and trade flourished across the continent. It was through this trade that new ideas about timber framing began appearing from France.

The subsequent advance in joint technology required not only greater skill but also better tools. Using a joint gave the advantage of connecting pieces of wood together without using other materials that could rot. This enabled the wall posts to be jointed into a cill (sill) beam or soleplate which ran around the building just above ground level. These free-standing box frames were initially simple in design and the use of A-frame rafters on the roof was unchanged. Unlike post-hole frames, which derived much of their stiffness from being half-buried in the ground, free-standing structures are liable to spread at wallplate level because of the weight of the roof. Framing technology

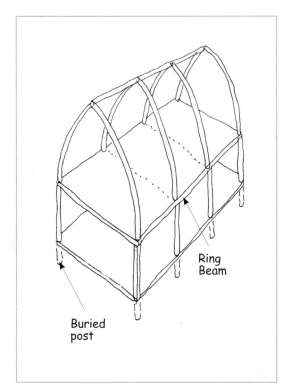

Ring Beam

Buried post

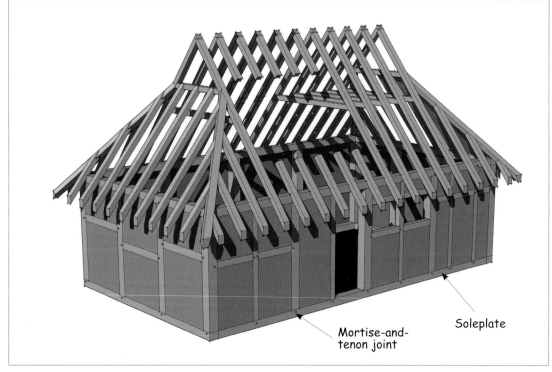

Mortise-and-tenon joint

Soleplate

quickly expanded during this period to cope with such problems, developing on a trial-and-error basis and leading to relatively sophisticated designs.

Soon tie beams were placed on the wallplates above the main posts, to stop the walls from spreading. Diagonal braces were introduced between the main beams and posts to stiffen up the frame and stop it from racking or swaying.

The roof proved to be another problem, as the individually paired rafters were liable to collapse like a stack of dominoes. Crown posts were therefore placed centrally on the tie beams to support a horizontal beam called a crown plate. The purpose of the crown plate was to support the collar – a horizontal timber jointed between the rafters. This transfers much of the roof load onto the tie beams and from there down the posts into the ground. Braces were introduced between the crown post and crown plate to stiffen the roof in a longitudinal direction. Braces were matched out of curved timber and the crown posts were often shaped and moulded to reflect the Gothic style of the day. Crown-post roofs were gradually superseded by purlin roofs, which were originally developed by the Romans and spread from the Mediterranean coast. A purlin is a horizontal beam that runs between the main cross frames, and the houses so constructed would have used many purlins to support a low-pitched roof that made them appear similar to Doric temples (such as the Parthenon at Athens). This design was then modified for the rainy British climate, by increasing the pitch to enable the use of thatch.

Far left **Primitive A-frames were eventually connected to a ring beam and posts buried in the ground. These types of post-hole structures were still used for peasant dwelling up to the nineteenth century**

Left **During the twelfth century mortise-and-tenon joints became established practice and posts were connected into a soleplate, instead of being buried in the ground**

Top right **Tie-beams and wall braces were introduced to stiffen the buildings, which previously had gained a lot of their rigidity from having posts buried in the ground. Crown-post roofs were developed to stop the rafters collapsing like a stack of dominoes**

Right **Purlin roofs, which came from technology developed by the Romans, eventually superseded crown-post roofs**

Racking

The unwanted sideways movement of the frame, causing the joints to deviate from their true angles.

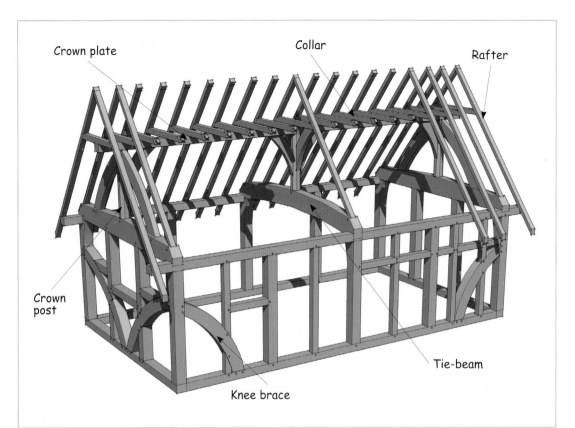

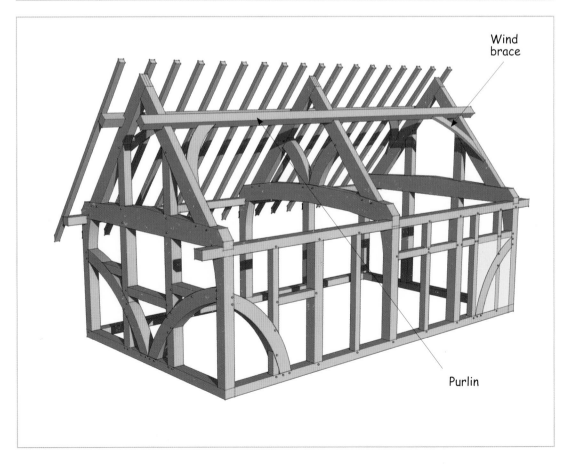

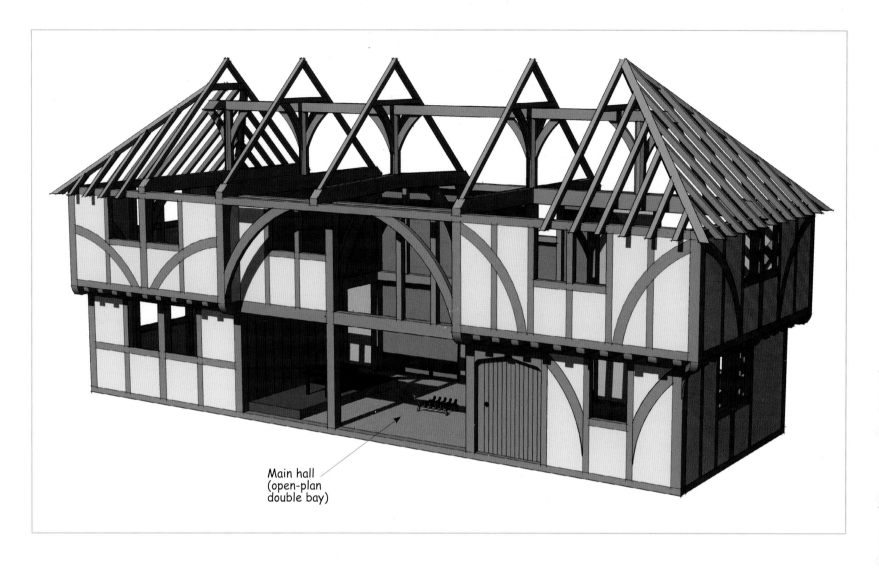

Main hall
(open-plan
double bay)

The Middle Ages

The Middle Ages heralded the golden age of timber framing in Britain. By this time much of the country was deforested but timber-framed buildings were still the preferred method of construction and their design had evolved into intricate structures, made up of many pieces of wood. The style of these buildings, to a certain extent, reflected the area where they were built, and the country was split roughly in two, with the South-West, Midlands, Wales and the North occupying the 'highland zone' and the South and East in the 'lowland zone'. Cruck frames (see page 47) were predominant in the highland zone, where large curved timbers (often a pair that was split from the same tree) rise from ground level to the ridge. The lowland zone was characterized by crown-post roofs and aisled buildings.

The pinnacle of British timber-framed buildings was the medieval hall house. Most surviving hall houses were made for the wealthy classes such as

landowners or merchants, and were centred round an open hall. The structure was divided into bays – in much the same way as a modern steel-frame building is today – that are formed by the natural divides in the frame where the major timbers are located. Each bay would logically form a room, and in the case of a hall house the two central bays would normally be left open.

The arrangement of the hall house had great social significance and they all followed the same basic form. The entrance to the building was through a door in the side wall which led into a screened passageway. On one side there would be doorways leading to the service accommodation, where the buttery and pantry were housed and where the servants slept. On the other side of the passageway you would enter through a screen into the main hall. The central hall would be divided into distinct zones.

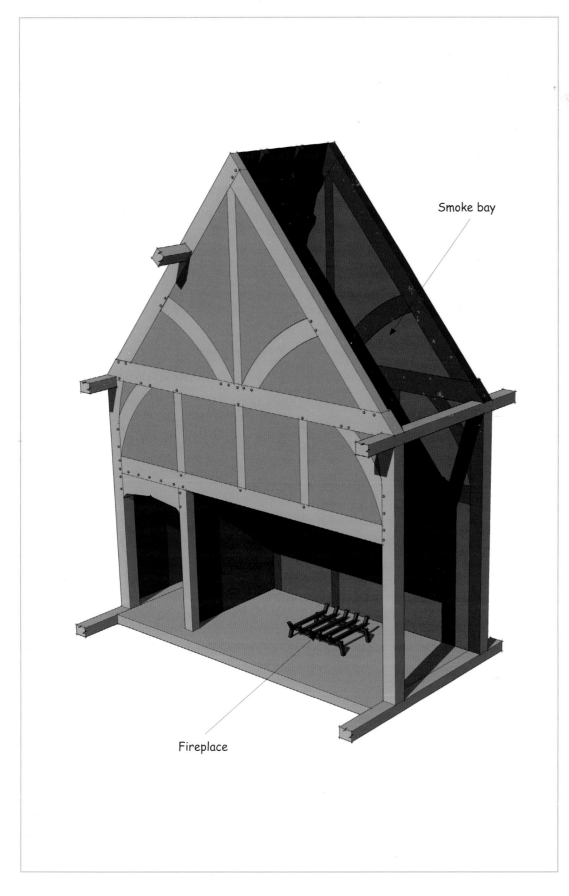

Smoke bay

Fireplace

An open fire would be situated on the ground near the centre of the two bays, and at the far end of the second, upper bay there would perhaps be a dais or raised platform. This is where the lord and his family would sit at the high table. There would have been no chimney, so the smoke from the fire would have risen into the roof space, and escaped from between the eaves or out through a purpose-made hole. Behind the high table there would be another division and a doorway leading to the 'solar', which was where the private sleeping quarters were housed.

It was quite common to add bays or floors on to hall houses at different times, rather like adding an extension to a modern house. The designs of the frames and especially the central truss in the hall varied between the different regions of the country, but common to all these buildings was the high standard of carpentry. This was revealed in the architectural details left by the carpenters, many of which would not even be seen. The carpenters who were making these frames were not only good craftsmen but the artisans of their day; their role was that of architect, engineer and surveyor in one. Their craft had reached a level of excellence that enabled them to express beauty in the structure through the natural materials they were using.

Left **Smoke bays replaced the open hearths of the hall houses. They were liable to catch on fire, so were later replaced by brick-built chimneys**

Post-Medieval Frames

The open-hall design gave way to a more common two-storey accommodation in the post-medieval era. These later designs normally had 'smoke bays', replacing the open hearth of the hall house. The smoke bay initially consisted of two cross frames, closely spaced, which would be open at the bottom and filled in higher up, with a hole in the roof. This was very dangerous and even though the insides would be rendered with lime mortar, they often caught fire. Later on brick chimneys were usually built inside the smoke bays to make them safer, and hall houses were converted by building a chimney next to the entrance or sometimes in the screen passageway itself, and floors were put in above the open hall.

Large amounts of timber were used to build houses during this period. More extravagant frames began to appear as a way of showing off one's wealth. Close studding, where vertical timbers are framed closely together in walls, became very chic, probably because it was expensive. During the seventeenth century the demands on oak became increasingly great. Large amounts of it were used for building ships and converted into charcoal for making iron. This led to higher prices and the import of cheaper softwoods. After the Great Fire of London in 1666, laws were passed that restricted the use of timber in buildings in the capital. Around the eighteenth century pattern books for carpenters were in circulation that detailed the popular designs of the day, especially new types of roof trusses from Europe, such as the king-post truss, which could span large distances. Brick houses became fashionable and the carpentry element was downgraded to the roof and internal floors. Iron bolts began replacing mortise-and-tenon joints, and the skills and knowledge of the great timber-framing age gradually diminished.

Throughout its evolution a pattern of English framing developed so even if there were regional differences certain techniques became standard across the country. English and other European frames were generally made by a technique called 'scribe-rule' framing. This method assumes that no timber is square or straight, so each piece has to be carefully laid on top of each other in a full-sized two-dimensional layout, before the position of the joints are scribed using a plumb bob (see Chapter 5 – Making Frames). These techniques were taken to the New World, where the European-style 'scribe-rule' framing was originally practised. By the late 1700s evidence shows that American frames underwent a transitional stage, where the framing members were roughly squared at the position of the joints, to speed up the process of scribing. By the early 1800s a new system was developed in North America called 'square-rule' framing. The key factor of this system is that it assumes that every out-of-square timber contains a mathematically square timber within it. Therefore if a dimensioned housing is cut in a post, square to a reference face, a connecting beam can be mathematically cut to joint into it perfectly. The practical advantages of this method meant that frames no longer had to be laid out in full, and therefore each member could be cut, for instance, in a shed during the cold winters. The abundance of straight timber at the time meant that curves were no longer used, so braces, for example, could be cut to be interchangeable within the frame. The speed and simplicity of this method meant that it superseded the scribe-rule method and it became the standard method of construction throughout North America, until it was replaced by balloon framing using small-section timber in the 1900s.

Right **Square-rule framing involves the assumption that each timber has a mathematically square timber within it. This method creates more uniform frames than the scribe-rule method that this book concentrates on**

The twentieth century saw massive changes in materials used to construct houses. Out went the use of traditional lime and in came concrete. Roofs were constructed out of 'gang-nailed truss rafters' (most still are), which were made from imported softwood, and metal window frames gave way to UPVC ones.

Perhaps as a reaction to these changes, the 1970s saw a revival in the art of not only timber framing but many other vernacular crafts as well. This was started in the USA, by the likes of Tedd Benson and Jack Sobon, and by the 1980s the revival had spread to the UK. Since the mid-1980s traditional British-style oak framing has grown in popularity and is now a well-established construction method again. It has even taken hold in the USA and Canada, where traditional British methods are becoming increasingly fashionable. There are many reasons why traditional oak framing has become so popular: perhaps it is the trend for open-plan living, or the speed of site construction or the fact that it is environmentally friendly. I could go on, but when it comes down to it, most people I meet just love the smell and touch of the oak, as soon as they walk into the workshop.

Right **Green oak being hewn**

Chapter 2

Starting Off

For many people building their own home is one of life's great milestones, ranking alongside getting married or having a baby. The diversity of information now available on the subject has led to many self-builders dreaming of building a different type of house, one not constrained by the post-war building methods seen in so many housing developments. In this chapter I will examine the basic steps a 'self-builder' needs to take, from finding a plot (or in the USA, a lot) through to applying for planning permission. These steps are not exclusive to oak-framed buildings but nevertheless they are critical to the success of the whole project. Time spent ensuring that the budget, design and planning are correctly thought out will be well rewarded later on. Many self-builders rush through this initial stage, believing the actual building work to be the most important part of the project, but if the design is wrong in the first place, or the money is not in place to finish the project, it doesn't matter how good the build is.

I should point out at this stage that a self-builder is someone who commissions the design and the construction of a house for their own occupation. Because they plan to live in the property, it has not been built on a commercial basis. The extent of hands-on work a self-builder actually does will of course vary and depends on many factors, particularly their own skills. Many are not involved in the building work at all, preferring to leave it to professionals, but most have an input into the design. Perhaps self-building has become so popular because people increasingly want to live in an individual way and have some influence over the space they inhabit.

The Budget

The importance of the budget in any building project can't be over-emphasized, and it is surely any self-builder's starting point. For most people building their

dream house has to be practical and financially viable. The end cost of building the house should be less than its final value. Value is a subjective quantity, and for some, the final cost of a project only plays a part in the value they attach to the building. The end value of a building project is normally determined by the current market value of the house if it were put up for sale. As we all know, a property's market value can rise or fall, so you should allow a safety margin. However, the end value may be less important to you, especially if you are building a house to live in for a long time. In which case, while you may make a paper loss in the beginning, eventually, with rising house prices, your property may become worth more than your initial investment. For most people building within a budget is essential, and unless significant private funding is available, money for the project will have to be borrowed from institutions who will want to see a properly worked-out and well-presented business plan.

Left **Self-building can be an amazing achievement but good planning is critical to the success of the overall project**

How to Calculate a Building Budget

A total budget may be broken down into: the cost of the land, the building cost and the many incidental costs which accumulate such as architects' fees. Once you have worked out the total budget you can afford, deduct the land cost and incidental costs (assuming you know them), and you will be left with the building cost, which is the starting point for estimating the size of your project.

The Square-Metre Rate

This leads us on to thorny issue of basing the building budget on square-metre (m^2) rates. This can vary widely and is complicated by the variety of choices of materials, finishes and interior fittings, for example. At the best of times this can be difficult to arrive at accurately. The way to use a m^2 rate is first to work out the habitable area of your intended house and then multiply this by the m^2 rate figure to give the total estimated building cost. The first part of this equation seems simple enough, if you know the size of house you wish to build. There are many ways of estimating this but perhaps the best way is to look around at other houses in the area and find out how much accommodation is actually available in a house of a certain size. Just looking at how many bedrooms a house may have can be misleading. It's the second part of the equation (the m^2 rate) that can cause a budget to spiral out of control. It is common to be over-optimistic and underestimate how much it is going to cost per square metre to actually complete a build. This figure is based on many assumptions and should only be used as a guide. It is preferable to opt for a high initial figure, than to find out later on that you have seriously miscalculated your costs. The m^2 rate varies with market conditions, so refer to up-to-date publications such as self-build magazines, where costings are given for self-build projects. Asking your local architect or builder is another good source of information and, unlike the self-build magazines, their figures should reflect local conditions.

The m^2 rate is affected by many factors and you should consider the following:

⋀ **The area of the country in which your house will be built** – Labour rates can vary greatly with geographic location and it is worth checking at an early stage.

⋀ **The type of construction and quality of finish** – It is possible to double the costs, for example by using natural stone tiles on the roof, instead of pre-cast concrete ones.

⋀ **The quality of interior fittings** – This plays a larger role than you might think, for example having a handmade kitchen rather than an off-the-shelf flat-pack, or choosing underfloor heating instead of having conventional radiators.

⋀ **Site conditions of your intended building site** – These may not be apparent until the land has been selected. It is well known that sloping sites cost more to develop, but certain ground conditions require special foundations and this can add significantly to building costs.

⋀ **The risk of flooding** – It is worthwhile checking the risk of flooding in advance, as insurance and flood-prevention works can be costly and complicated. If you intend building on a flood plain, special pilings may be needed and these will increase your m^2 rate. (See www.environment-agency.gov.uk in the UK or www.floodsmart.gov in the USA for more information.)

⋀ **Obtaining services** – This may be a significant cost factor if your site is a long way from an electricity connection or similar services.

⋀ **The habitable space** – This not only comes into the equation when multiplying the m^2 rate, but it also affects the m^2 rate itself. It is usually assumed that both floors of a two-storey house will be usable floor area. However, it's reasonably common in an oak-framed building to have a void area in some part of it; for instance, there may be no upper floor above the downstairs sitting room (see right), much in the same way as a medieval hall house (see page 21). The m^2 rate should therefore be increased accordingly to make an allowance for the lost floor areas in a building of a particular volume.

Above **The final cost of the build will depend on a number of factors such as the quality of the finishes**

Left **When calculating the m² rate, void areas within the building also need to be accounted for**

When calculating an initial estimate for your building cost, it is preferable to err on the pessimistic side, especially as you are unlikely to know all the previously mentioned conditions from the start.

Incidental Expenses

Incidental expenses will continually occur throughout the project and must be accounted for at the beginning. It is tempting to overlook these, as self-builders naturally focus on the two big expenses: the cost of land and the cost of the build. However, incidental costs can make the difference between a healthy profit and a heavy loss.

At some point or other as a self-builder, it may well be necessary to commit yourself to the project without knowing the final costs involved, especially if you buy the land before the final design has been agreed. In areas where there are very few potential sites and demand for them is high, many people are prepared to pay over the odds to secure the land they want, which can result in final selling prices that exceed a third of the final value of the property.

If the actual land cost is more expensive than you initially thought, savings may be made on the build cost, if considered carefully at the planning stage. The more accurate and detailed the original build and project plan is, the more accurately costs may be analysed and clear judgements made about possible savings.

Finding Land

The crux of any building project is finding the right land to build it on: not only that, but finding a plot (lot) in the right location and for the right price. Not an easy thing to do in today's increasingly crowded world! Searching for the right site is something of an art form in itself, and inevitably involves a great deal of hard work, a fair amount of luck and some compromise along the way.

Location, Location, Location

The starting point has to be the location, as it is going to directly affect your quality of life, the design of your dream home and its final value. There is no point in building your house in a beautiful but remote part of the country if it is going to be impossible to get to work on a Monday morning, or by a river that looks beautiful on a calm summer's day but is prone to flooding during winter.

Buy a Map

A good place to begin your search is by purchasing a large-scale map of the area in which you wish to live. This allows you to pinpoint empty buildings or identify a region in which development is more likely to be approved – for example, where it falls within the pattern of existing conurbations. Study the map carefully and drive round the area you are interested in. Ask local people if they know of any land for sale or anyone who might be interested in parting with a piece of land. This might sound like

Below **The site of an oak-framed building will have important implications for the cost of building it**

The main incidental costs that occur are:

∧ **Taxes** – Those paid on land (such as stamp duty in the UK) will not be recouped; however, there are schemes by which self-builders can recover tax paid out during the construction process, and it is important to establish the current regulations at an early stage in the proceedings. The Inland Revenue (www.hmrc.gov.uk) in the UK and the Internal Revenue Service in the USA (www.irs.gov) are good places to start looking for information.

∧ **Solicitors' fees** – These will be incurred during the conveyance for transferring the money during the purchase, checking the title of the property, doing local searches and so on.

∧ **Architects' fees** – Normally based on the estimated building cost. Other professional fees, such as for engineers and quantity surveyors, may also have to be factored in.

∧ **Submission fees** – Payable for the submission of plans to the local authority for approval.

∧ **Accommodation** – This can be a costly expense during the construction phase but money can be saved if you are prepared to buy a caravan (trailer) and live on site. Permission is not usually required for such a temporary structure, although it is worth checking, and they normally have to be removed once the construction has finished.

∧ **Building loans** – The interest payments on any finance agreements must be allowed for and it is advisable to allow for the possibility of an interest-rate rise during the build if you have agreed a variable-rate loan.

∧ **Site insurance** – An important element of any build that should be arranged to commence as soon as the site is purchased. This usually covers public liability, employer's liability and contract-works insurance.

∧ **Land clearance** – This could include demolishing an existing building. It can turn out to be costly if there are toxic materials to be removed, such as asbestos.

∧ **Finishing works** – A cost that is often ignored but covers such important and sometimes expensive factors as landscaping, driveways and outbuildings.

∧ **Warranty schemes** – These are often a necessity if you wish to borrow money, and offer financial protection in case the build suffers problems. They may also cover a period after the build's completion. Warranty schemes usually involve a series of inspections to ensure the workmanship and materials are up to standard, and there are several different types of warranties available. Building insurances vary in the USA from state to state, so check locally to see what is available.

Above **Choosing the right designer for your project is one of the most important decisions of any building project**

areas where building is likely to be restricted. The local plan also highlights areas that are designated as flood plains, infill, conservation areas and so on. The maps can sometimes be downloaded directly from the local authority's website. If land does not lie in an area designated for development then there is no point pursuing it.

The Agents

Estate agencies (real-estate agents in the US and Canada) generally fall into two categories: town and rural. Obviously it is a good idea to register with all of those covering your chosen area. Town agents tend to have a high turnover of properties and are more open about what they have on their books at any one time; however, they are less likely to sell undeveloped land. Rural agents are more likely to have land available for development, and many actually have departments that specialize in obtaining planning permission for their clients. Unfortunately, agents are quite often loath to sell to self-builders, preferring instead to deal direct with developers, allowing themselves two bites of the apple: once when they sell the land and once when they sell the finished property. The trick is to constantly pester them, so they know you are serious, and widen your search to include not only vacant land but also old buildings that could be demolished to provide a suitable site.

The advantage of a 'knock down' job over constructing on undeveloped land is that the essential services are usually in place. You may be able to live in the old house while you are building the new one, and you might be able to save some money by recycling some of the materials, but you will have to remove the old property at some stage. There is a fairly good stock of these buildings available, often in great locations where planning permission wouldn't be granted today. If you do decide to go down this route, it is very important to check with the planners what, if any, development would be possible before buying, and then make an offer subject to getting planning permission. In that way, if planning permission is refused you won't be left with a clapped out old bungalow on your hands. Sometimes these properties come up for sale with planning permission to build a new house on the site, but the price usually reflects this. If this is the case, and the design is not what you want, don't despair. It usually is much easier to change the design and reapply for planning permission once a precedent has already been set, so long as the new application does not vary a great deal in size or shape from the original.

a long shot, but you would be surprised how often it throws up results. It is always possible that you might find a wedge of redundant land between two buildings that has not been built on yet. Finding the title-holder of redundant land may prove difficult. If local searching and questioning turns up nothing, then try www.landregistry.gov.uk in the UK or www.uslandrecords.com in the USA.

Look at the Local Development Documents

To check whether planning permission is likely to be given in the UK, contact your local planning department and ask to see a copy of the Local Development Documents which show in detail where development is permitted. In the US you should contact your local authority to check the zoning laws to find out what type of development is permitted. The plans contain detailed policies and detailed maps of land classifications, including

Another type of agent that is worth considering is a land-finding agency. They normally advertise in the back of self-build magazines, on-line and at home-building shows. For a small fee you can subscribe to their services. In return they will send you every plot (lot) that they have on their books on a monthly basis, for a given area. This can turn up a huge number of potential sites, but because of the way the data is collected, the information is often out of date. Despite that, vendors occasionally have more than one plot (lot) in a similar location, so even if the site you are interested in has already been sold you may have gained a useful lead.

Development Policies

These govern the use of land for development and the protection of the environment, and are the most significant consideration when applying for planning permission. The National Planning Policy Framework (NPPF) was brought in during 2012 and deals with planning on a national level. On a neighbourhood level local authorities work to their own Local Plans, which cover the strategic planning and development management policies (called the Core Strategy) of each region. Planning applications have to be determined in accordance with the Local Plan policies and Neighbourhood Development Plans but there are also other documents that they take into account. Sustainable Development is a key factor in getting approval more quickly.

In the US the zoning system controls development. Zoning laws vary from state to state, but normally specify industrial, residential, commercial or recreational areas. Apart from controlling the use of land and buildings, zoning laws also may regulate the size of buildings and their lots, the concentration of any development, and the use of the land. Some even protect buildings and areas of historical significance. Zoning laws are not necessarily permanent and sometimes they are relaxed.

The Initial Design

Finding land is one thing, building the house of your dreams on it is another. Unless you have a flair for design, and know about the planning laws, you will probably wish to consult a professional before purchasing the land. The advice will have a lot to do with whether the land has outline or full planning permission. If it has outline planning permission you will need to check what conditions have been made, and if it has full planning permission, whether you can live with the existing design or want to change it. If you do want to change it, unless it is only a minor alteration, a full new planning application will have to be made. In the US 'sealed' building plans of the final design have to be submitted by a qualified architect or engineer to the local government, and then passed before any building can commence.

Choosing a Designer

'Who', you ask, 'is going to design my beautiful new oak-framed house?' Well, it is an important decision, and not one to be taken lightly. Whoever you choose is going to have a substantial effect on the whole project, for better or worse. The design they produce, and the one you hopefully get planning permission for, will be pivotal to the whole build. Mistakes made at this stage can have a knock-on effect throughout the whole project. Although many building problems can be solved 'on the job', basic design oversights usually can't. Designing a house is a very skilled process, and therefore, it follows that the designer needs to be very accomplished at the task, and have a flair for the process as well. Unusually, amongst professions, house designers have to be scientific in their understanding of structure, building techniques and budgetary requirements, as well as creative and artistic in the application of that scientific knowledge.

House designers fall into several categories, although the distinction between them can sometimes seem blurred. It is important to check their qualifications and background, as many people describe themselves as architects without being suitably qualified. The main groups are listed below.

In the end, whoever you choose needs to be able to translate your ideas into reality. There needs to be a very good working relationship between designer and client, and the success of the project relies on it.

Those involved in the design of oak-framed buildings include:

⋀ **Architects** – The title of Architect is protected by law and only those registered with a professional body having undergone the relevant training are able to use it. Many architects are also members of bodies that lay down very strict guidelines governing professionalism. They must also have professional-indemnity insurance, in case they are sued for negligence. The US has a similar system and only qualified architects are allowed to use a 'seal' on drawings submitted for Code Compliance. Architects are trained in design and as well as concentrating on the specifics of a project they also have the ability to look at the big picture. They should have an artistic flair as well as technical know-how, but quality varies and it is important to view examples of your selected architect's work before you appoint him or her.

⋀ **Architectural technologists/technicians** – Like architects, architectural technologists and architectural technicians have to undergo training to become qualified and become members of a professional body. Their primary focus is on the technical side, rather than the design of buildings, although many are very good and successful at it, and they often work with architects, for example to produce the building-regulation drawings from the architect's original planning drawings.

⋀ **Building surveyors** – Chartered Surveyors design houses and are regulated by their professional body, the Royal Institute of Chartered Surveyors (RICS). They have many members worldwide, covering all manner of interests from real estate to environmental issues, but the ones we are most interested in are building surveyors, who concentrate on construction.

⋀ **Non-qualified designers** – By non-qualified I mean those who have not undertaken any professional training in building design, but are, perhaps, qualified by experience. Many of the most competent but also some of the most inept people I have worked with fall into this category, and after all, it is not a legal requirement to have a qualification to design a house, but it helps.

The Client's Brief

OK, so you have found a location and selected your designer, now designing your house needs to start in earnest. But where do you begin? Good design should be an evolving process, which draws on information from many different sources. There will be many important considerations to be included, ranging from how many bedrooms are required, to where the cat-flap is fitted! To make sense of all this information, a process needs to be laid down by which it can be properly organized into a format that can help achieve the final design. The following list contains a few pointers to aid the design process:

∧ **Make a list** – Write down all of the accommodation that you are likely to need, and try to include everything you can think of. There will be the usual rooms, such as the bedrooms and the kitchen, but also any special rooms that you feel you can't live without, like a darkroom or a sauna. Think about the exterior of the property: do you want a detached garage, or have you got a yearning for a reed-bed sewage system? For each room think about how you will use it, what will go in it and any special requirements. Remember practical things like storage space and security, as designers often overlook these.

∧ **Site map** – A detailed site map is a priceless tool and if you haven't already got one then they can be obtained from the Ordnance Survey in the UK (www.ordnancesurvey.co.uk) or the United States Geological Survey in the USA (www.usgs. gov). A variety of maps are available with some scales as large as 1:500. Once you have a detailed map, it is possible to plan out your site. Mark on the position of the access and any other useful features, such as the main services, large trees or overlooking neighbours. (Check with the local authorities whether trees are subject to a tree preservation order.) Whilst these plans are usually accurate, I would always recommend having a proper survey carried out by a chartered surveyor.

∧ **Site orientation** – Orientation is all about sunlight and views and how best to maximize these. Mark on your map the position of sunrise and sunset, and if possible show these positions for winter- and summertime conditions, and also at different times of the day, as this may dictate the orientation of certain rooms. At the same time, mark the direction, or directions, of the best views, and make a note of any features already in place that may obstruct these views. Also note the direction of the prevailing weather as well as any places where you may be overlooked by your neighbours.

Left Site orientation needs to be carefully planned. A large-scale map should be purchased and marked up with any salient information, such as the position of the sun and the prevailing weather

Building location – Armed with your detailed map, showing every piece of site information that could affect the design, the location of the building needs to be added. This is perhaps one of the most important parts of design, for which there is no written formula. The position needs to encompass all the information that you have gained about the site and also comply with any regulations. Firstly you need to think about the access and physical constraints. There is no point positioning a building that is impossible to reach, or is sited over a protected tree. You must then consider natural light and views, and how to gain the greatest benefit from these. And finally, consider how the boundary of the site interacts with the building and whether it is better to set it against the edge to maximize the garden space or to place it somewhere near the middle for landscaping reasons. The choice varies with every single site and has to 'feel' right and balance all the competing factors. That balance is the invisible element present in every good design.

Room location – Earlier in this chapter you made a list of the accommodation required. This can now be added to your map, with relevance to position and orientation. Simple bubble diagrams, such as the one to the right, are very effective for this purpose, and allow one to easily explore different ideas. Think about how the separate rooms are going to be used at different times of the day. The main living space will want to be positioned to have the best views and light – perhaps on the south side – whereas the spare bedroom or utility room could be positioned on the north side or where there is no view.

By the end of the process, you should have a fairly good idea of what you want your building to achieve within the location, and its spatial requirements. The next stage of the design is to incorporate your ideas within an oak-framed structure.

Below **A simple bubble diagram is a useful way of planning room locations**

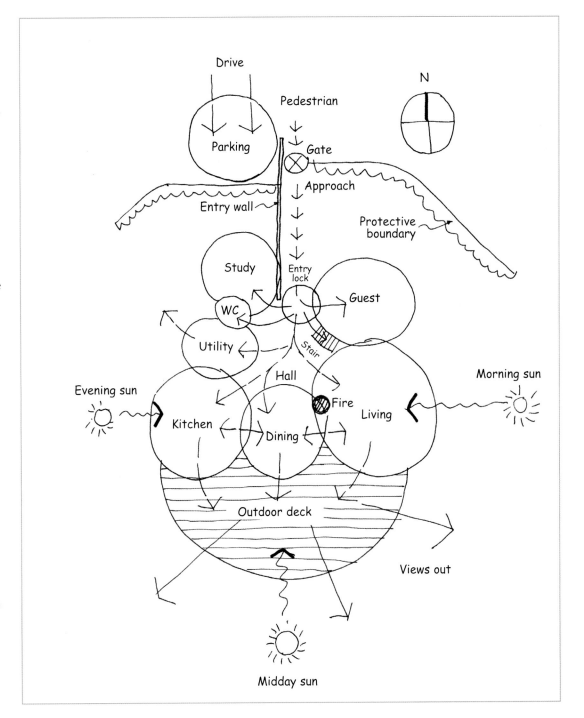

How Structure Affects Design

The famous architect Le Corbusier coined the expression: 'Form follows function' and whilst this is usually the case, with an oak-framed structure the design of the structure will dictate, to a large extent, the final design and spatial layout of the house. The following chapter goes into much greater depth about frame design and structural issues but for now it is important to know the basic principles of structural oak framing.

Oak-framed buildings are constructed by connecting a number of separate pieces of timber in order to form a skeletal frame that, unlike brick or masonry, does not rely on gravity and dead weight to hold it together. The oak beams and posts are fixed together using traditional joints, and braces are introduced between them to keep the whole structure rigid. The building is separated into bays by the cross frames, which transfer the load of the roof down the posts into the ground. The width of the bays and the span of the cross frames are determined by the size of timber available and tend to fall between certain limits. This means that a typical building may require five cross frames to support the roof structure, and therefore will be divided into four bays. In this way we begin to build up a grid system for setting out the frame, and it is this that will determine ultimately how the rooms in the house are laid out.

If designed correctly, structural oak frames require no additional load-bearing walls to support the floors or roof. So in theory, the whole structure could be safely left open from the ground-floor slab right up to the rafters, and in some cases it is. Generally, though, most frames are divided to some extent, and the best place for a room division is on the gridline between bays. The oak frame is the soul of the building and it expresses the structure wherever you look. When dividing it into separate rooms try to maintain this appearance by putting the divisions on the cross frames, so the structure is visible on both sides of the room. Occasionally, I have built a wonderful frame, only to find that the client has come along and divided it up into little boxlike rooms with no respect for the frame layout. The inevitable outcome of this is a hotchpotch of walls and timber everywhere.

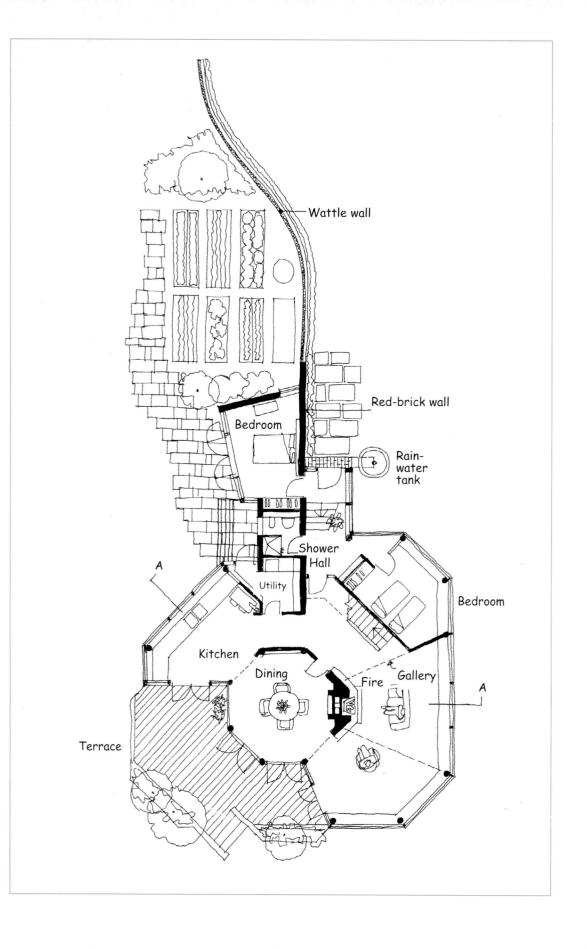

Above **Oak-framed buildings are ideal structures to create open-plan living areas**

Left **The bubble diagram from page 35 is turned into a house design. The layout of the frame is based on a grid system that determines the positions of the rooms**

Taking the grid system on board, you can begin to marry the room layout with a basic frame design in a complementary, rather than competing, fashion. Some parts of the house may need to be divided for privacy, such as the bedrooms and bathrooms, and other more communal parts left open. In the past people have tended to separate all the rooms in a house with walls, but in modern times the trend has moved more towards open-plan living. The kitchen, once relegated to the smallest back room, has become central in

our everyday lives and is now often connected with the living space. This can work perfectly with an oak frame, with the kitchen perhaps located in one bay and the living space in another, but with no dividing wall, and only the cross frame to identify the transition between these separate areas.

Planning Drawings

There comes a point in any design when you need to sit down with your architect and thrash out your ideas to produce a design. Armed with all the previous information in this chapter, you should have a fairly good idea of what you wish to include and where you would like it situated. Now the process of change and compromise will begin! How do you fit all those rooms into such a small space? Perhaps your outline planning permission only allows for a 2,000ft² (185m²) building but you will need at least 4,000ft² (370m²) to come close to your initial design, not to mention staying within the budget. Your architect should help to sort out many of these problems and counsel you to the best course of action. But beware: do not let other people ignore your thoughts and input, after all it is your house.

If you or your architects have no experience of oak structures, then it would be worth contacting an oak frame building company which can be involved with the design at an early stage. From the beginning there should be a close relationship between the house designer and the craftspeople who are going to make it. Each company will have in-house frame designers who will be able to advise on the best way of making the structure work and on how to achieve the most from your oak frame.

Submitting Plans

Hopefully, at the end of this process, you should have a sketch design that you, or your architect, can show the local planning department. It is always a good idea to make an informal approach to the planning officer ('building official' in the US) at an early stage. The last thing you want is to go through the trouble and expense of preparing a full set of plans, only to find out two months later that the planners would never have considered a building of that size or type in the first place. It is advisable to show as much detail as possible from the outset and to get this approved in principle to save any misunderstanding later on. This is where an experienced architect or designer can make all the difference, especially if they are local to the area and know the planners.

Once you have received an indication from the local authority that your designs are acceptable in principle, your final application can be prepared including your location and site plans as well as floor plans and elevation drawings. This will have to be

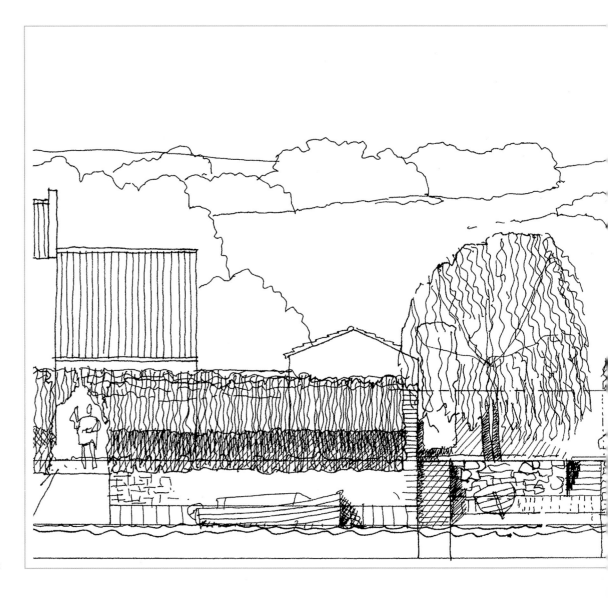

Right **Once the design has been finalized, planning drawings are submitted to the local planning department**

accompanied by information on the materials that are proposed for construction and any other relevant information, such as trees to be felled. All of this will have to be sent off with up to six copies and, of course, a fee!

Often during the course of a planning application it is necessary to adapt or make changes to the design along the way, so be prepared to make some concessions. If all goes to plan, the application will be granted permission by the planning department, within the guideline time period. Unfortunately, this often goes awry, and in the UK most new buildings are referred to the planning committee for approval. This is normally just a case of 'rubber stamping' the planning department's decision. However, I once built a house where the planning department were dead set against

giving permission, for a whole variety of reasons. The local planning committee decided to pay a site visit, and when they saw the plans and found out that the building was an oak frame, made by local workers, they decided to overturn the recommendation and permission was subsequently granted. So, if your planning permission is recommended for refusal, try lobbying those with influence over the final decision, emphasizing the positive points of your design such as using local craftspeople and environmental credentials.

Appeals

If the planning authority refuses permission or imposes conditions, it should give reasons for doing so. If you are unhappy or unclear about the reasons for refusal or the conditions imposed, talk to staff at

the planning department. Ask them if changing your plans might make a difference. If your application has been refused, you may be able to submit another application with modified plans free of charge within a certain period of time after the decision on your first application. If you think the decision is unreasonable, you can appeal to a number of authorities. In the US this varies from state to state, while in the UK Secretary-of-State appeals must be made within three months of the date of the council's notice of decision. You can also appeal if the council does not issue a decision within eight weeks. Appeals are intended as a last resort and they can take several months to decide. If you are considering an appeal it is wise to talk to a planning consultant who will advise on the likelihood of the

success of your appeal and can then act on your behalf. Planning consultants are usually ex-local-authority planners and understand the complexities of the policies and processes involved.

On-line Help

For further information on planning and for detailed guidance on how to apply for planning permission in the UK visit www.planningportal.gov.uk or in the UK and the USA approach your local authority – most of whom offer on-line services at the relevant address – to make and view planning applications on-line. Their websites can be a very useful source of information and can even can be helpful in your search for a potential site for your build.

Below **The beauty of using natural materials in an oak-framed building is that they blend almost seamlessly with rural surroundings**

Starting Off

⋀ **The budget** – Work out the total budget that you can afford and then begin to estimate the costs that are involved in the project. Calculate the land costs and the incidental costs such as agents' fees and subtract them from the total budget; this will leave you with the amount that you can afford to spend on the building. This figure is the starting point for estimating the size of your project.

⋀ **Finding land** – Not only must the land that you choose be in the right location, but it must fall within budget and be in a position that will enable your project to comply with any local development policies.

⋀ **The initial design** – This stage involves choosing a designer and laying down a clear brief of what you want. This should include what you want from the house in terms of living and storage space, the location and orientation of the building and the positioning of the rooms.

⋀ **How structure affects design** – Form follows function, so lay out a grid system to marry the design of the frame to the eventual layout of the house; this should create a cohesive building that will be both functional and show the oak frame in its best light. Planning regulations will also place restrictions on the design of the building, so take these into consideration before you begin.

From Tragic Beginnings to a Triumphant End!

In October 2000, after sustained heavy rainfall, a river situated in the south-west of England burst its banks. The river was in a flood plain, so some winter flooding was expected, but this year it was exceptional. The river rose by over 10ft (3m) and vast torrents of water gushed down the valley.

Top and above **The morning after the flood, as the waters were subsiding, the owners of the property discovered the wreckage caused by it**

Left **The new house was built on a concrete platform to the Environment Agency's guidelines, so that it would withstand any future flooding**

Above **A roof truss for the new frame being craned in over the river**

Above right **Planning permission for the replacement dwelling was recommended for refusal by the planning officer dealing with the case, but this decision was overturned by the planning committee after they paid the site a visit**

A couple in the area had a lovely wooden chalet, which was situated close to the river but high enough above it for flooding not normally to cause a problem. This year was different though and during one night the flood water started entering their house. The water rose to over 3ft (1m) inside the property and caused substantial damage. In the morning they were distraught to discover the wreckage that the flood had made of their chalet.

Once the flood had subsided the chalet was inspected by an insurance company, who declared it uninhabitable, and wrote it off for insurance purposes. The couple then approached their local planning department to enquire about building a replacement property. They were soon horrified by the planning officer, who informed them that their house was not a house at all but was classified as a mobile home. This meant that it would be extremely difficult, if not impossible, to get planning permission for a new property on their plot. They then spent the next year fighting the planning department for a Certificate of Lawful Use on the property, which was eventually granted.

This is where my friend the architect came into the story. He proceeded to design the couple a beautiful oak-framed house. It was a single storey building with a room in the roof and because of the risk of future flooding it was designed to be built off a raised foundation to the Environment Agency's guidelines. The couple showed the plans to their neighbours who all loved it, and so they were submitted for formal planning permission. Unfortunately the case officer dealing with the submission was the same person who had said their previous house was a mobile home. This planning officer subsequently recommended the plans for refusal. Normally when this happens, the Planning Committee takes the advice of the officer and will refuse planning permission. In this case though, they were persuaded to pay a site visit. After hearing that the new house was going to be made by local craftspeople in a traditional way, they unanimously approved the planning permission. Ever since the couple have lived happily in their beautiful oak-framed house, which they love. It just goes to show that 'it ain't over until the committee sings!'

Above **An aerial view of a large semi-aisled frame under construction**

Right **Wealden-style houses, like this one, tend to have jettied floors and large curved braces in the walls. The crown-post roof is just visible within the structure**

Chapter 3

Designing Frames

The design of oak frames has been evolving for many centuries and was primarily based on a system of trial and error. Put bluntly, if a beam was not sized properly, it would fail. The next time a frame was made a larger beam would be used until the design stood the test of time. This was not only true for the sizing of beams, but also for the design of frames and the correct use of braces. The timber available was another major factor in the evolution of the design of oak frames, as indeed it is now. As the span of a frame increases, so does the necessary section size of the main beams carrying the loads. Past a certain point these section sizes become impractical. What do I mean by impractical? Well for a start, it might not be physically possible to find an oak tree of the right dimensions to produce such large beams. Even then if you do find one it may produce a knock-on effect on the rest of the joinery in the frame. So, for instance, if you increase the size of a floor beam the post that it is jointed into will also have to be increased to accommodate the greater dead load and joint geometry. This in turn will mean the wall and soleplates will have to be made wider, and so on.

In the past, as now, oak frames were made by carpenters who had to be able to handle the timbers while the frames were being made, as well as raise them into position on site. The raising is mainly done by cranes these days, but there is still a lot of manual handling involved so ease of construction still plays an important role in the development of the design. Of course there are exceptions to every rule: examples which spring to mind would be Westminster Hall or York Cathedral. In these cases, they used exceptional pieces of timber and probably hundreds of carpenters to construct them, unlike the majority of timber houses built over the last millennium.

Bays

The diagram overleaf shows a typical frame, which can be broken down into four main planes, in other words the roof frame, wall frame, cross frame and floor frame. Each section of oak is connected together, using traditional joinery, in order to make a skeletal frame. This transfers loads from the roof, walls and floors down to the foundations. In order to transmit these loads effectively, principal framing members are located together to form a cross frame. This ensures that instead of having load-bearing posts and beams scattered randomly throughout the frame, they can be organized into clear groups.

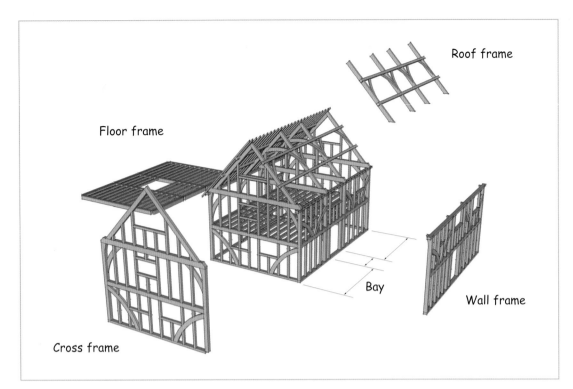

Floor frame

Roof frame

Bay

Wall frame

Cross frame

Left **This exploded view of a frame clearly shows the various planes that make up an oak-framed building. Each two-dimensional plane is made separately in the workshop. Members that are common to more than one plane are called primary timbers and those common to only one plane are called secondary timbers**

Below **Principal framing members are located together so the loads from the floor and roof can be transferred efficiently to the ground. Cross frames support the purlins in the roof and the joists in the floor. The gap between each cross frame is called a bay. Bay sizes can vary and offer natural divisions within the building**

Horizontal structural timbers in the frame carry the loads from the roof and floor to these cross frames. If a member is common to more than one plane, it is called a primary timber. A good example of this would be a principal rafter, which is common to the cross-frame plane and the roof plane. Those in only one plane are called secondary timbers.

The gaps between cross frames are called bays, these form natural spaces within the building. The use of bays is very common and can be seen in a great deal of modern construction. Next time you enter a supermarket, check where the main posts are, and you will see how the structure is divided into bays. In the past, houses were commonly described in terms of how many bays they had, in just the same way as people today might describe the size of a house by how many bedrooms it has.

Width of Bays

Bay sizes vary but normally they are spaced apart by between 10 and 13ft (3–4m). If the bays are much wider than 13ft (4m), the horizontal timbers connecting on to the cross frames, such as the purlins and floor joists, become excessively large in order to carry the required loads. These loads, such as the weight of the roof tiles or people and furniture on the floors, cause deflections in the horizontal beams. The deflection increases in proportion to the cube of a beam's length for a given load. If the bay size is too small, the frame looks overcrowded by the

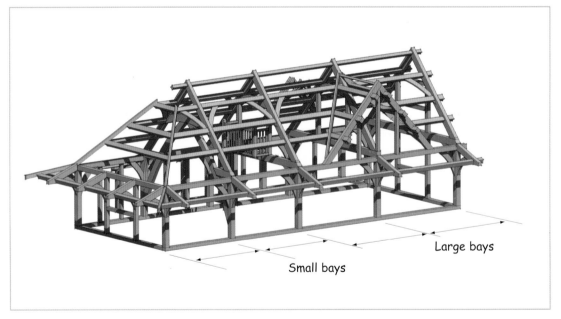

Small bays

Large bays

close proximity of the cross frames to one another. Hence it is important to achieve the correct spacing for both the structure and aesthetics.

Bays offer a natural division between rooms in the house. So for instance, one bay may contain a medium-sized living room or a bay split in half could contain two bedrooms. The bays in a house do not all need to be the same size, and in most cases, it is preferable to have different bay sizes in various parts

of the house. If your house is 15m long, for instance, you could have four equal bays of 3.75m or two bays of 4m and two bays of 3.5m. (Alternatively, if you prefer imperial measurements, a 48ft long house could be divided into four equal bays of 12ft or two bays of 14ft and two bays of 10ft.) If it is a two-storey house, not all of the bays need to have a first floor, so for instance the living room, situated in perhaps two bays, could be made open to the roof.

Cross Frames

There are three main types of cross frame: truss and post, cruck and aisled construction, and their various uses depend on a number of design and structural considerations (and also on the availability of suitable timber)! The following list to the right is a summary of each type:

Right **Truss and post**

Below right **A detail of an aisled-construction cross frame**

Below **An entire aisled-construction frame**

Bottom **Cruck frame**

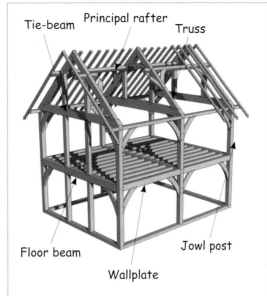

Tie-beam · Principal rafter · Truss · Jowl post · Wallplate · Floor beam

Aisles

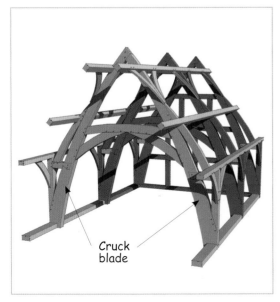

Cruck blade

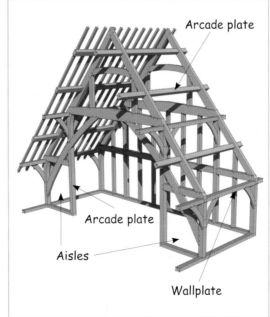

Arcade plate · Arcade plate · Aisles · Wallplate

Catslide Roof

A catslide roof is a long sloping roof (usually at the rear of a house) that has a continuous pitch from the apex of the roof to its eaves which are normally adjacent to the ground floor ceiling. These are particularly common in New England, USA, but are also seen elsewhere and are known as saltbox roofs in the southern USA.

∧ **Truss and post** – The primary roofing timbers, which are triangulated, make up a truss. There are many different types of roof truss, but fundamentally they consist of large principal rafters, usually 10x6in (250x150mm), jointed into a tie beam. The trusses take the load from the purlins – the longitudinal members in the roof – which in turn carry the load from the common rafters. The truss sits on top of a wallplate which is supported below by the main posts, or jowl posts. If the building is more than one storey high, it is probable that the jowl posts would also have a floor beam jointed into them. The floor beams, like the tie beams, are likely to be the largest timbers in the frame and typically are at least 8x12in (200x300mm). The width of the cross frame is determined by the size and availability of these timbers, and although it is possible to obtain oak beams of up to and even over 26ft (8m) long, their necessary section size would be so large that a 20ft (6m) cross frame width is the most practical maximum to set.

∧ **Aisled construction** – When a greater span is required, then it is better to switch to an aisled frame. These normally consist of two aisles running the length of the building; effectively a lean-to added on either side of a truss-and-post frame. The internal posts become arcade posts (or aisle posts) and the internal wallplate becomes the arcade plate. Aisled frames can be arranged in many different ways, which makes them a very useful structure when designing a large building. The aisle can be designed so that the pitch of rafters produces a 'catslide' roof, or even be of different spans. Aisled buildings do not need to be symmetrical, and many are built with only one aisle. The width of the aisle is generated by the pitch of the roof and the height of the wall and arcade plates.

∧ **Cruck frame** – Essentially, cruck frames are pairs of curved timbers that reach from the floor to the apex of the roof. Usually, a pair is made by splitting a large curved tree in half to give two symmetrical blades. The cruck blade does the same job as the principal rafter and jowl post, and naturally forms a structural triangle. Because of the difficulty in sourcing large cruck blades, many crucks are produced by jointing straight pieces of timber somewhere along their length.

There are three basic types of load acting on an oak frame structure.

/\ **Dead loads** – These are the weights of the structure itself and all other permanently attached materials such as roof tiles and wall coverings.

/\ **Live loads** – These are gravity loads which vary in size and location and would include the people and furniture that occupy the building and also snow loads.

/\ **Wind loads** – These are variable loads which depend on the aerodynamic behaviour and location of the building. These loads are dynamic because the of the way the wind constantly changes speed and direction.

Loading affects the frame in many different ways, and understanding these forces will lead to better structural design, using the correct size and grade of oak and making the right joinery decisions. Outlined below are a few pointers to consider when designing the frame.

/\ **A sustained load** – These can cause creep deflection (permanent bowing) – in the same way that an overloaded bookshelf will eventually bow and remain so, even after the books have been removed – so correct sizing of timbers under heavy load is important.

/\ **Load duration** – Timber is capable of carrying greater loads over a short period of time than over a long period. This is helpful when considering occasional live loading such as wind and snow.

/\ **Purlin loading** – Purlins are likely to carry the greatest load whilst they are still green and can be subject to all three types of loading at the same time. Great care needs to be applied when selecting the timber and processing the joinery.

/\ **Floor joist loads** – Floor joists should be carefully selected because they can be subject to some unusual live loads, such as a group of teenagers having a party!

/\ **Braces and wind loading** – Braces carry a large amount of force generated by wind loading, which is transferred through the walls and roof into the cross frames. Braces should always be placed in pairs where possible, so that when one is in tension the opposite one is in compression.

/\ **Beam sizing** – This needs to take into account any jointing requirements.

Jettied Beam

A beam that supports an upper part of a building that projects over the lower part, for example a beam that supports a protruding bay window on the first floor.

Stress is defined as the force acting on a member per unit area. There are four fundamental types of stress that we need to understand when it comes to designing oak frames, namely: compression, tension, shear and bending.

/\ **Compression** – This tends to compress or crush a member. If, for instance, a beam is supported by a post, the joint between them is in compression. Most structural joints in the frame should be designed to act in compression, as they tend to push together rather than pull apart.

/\ **Tension** – This tends to pull or stretch a member. Members in certain types of open trusses are in tension, for instance the collar of an arch-braced truss. As the feet of the truss try to spread, the joint between the principal rafter and the collar is under tension. This tension needs to be resisted by using careful joinery, such as a dovetailed joint.

/\ **Shear** – This results from two equal forces acting in opposite directions. There are two types: vertical and horizontal shear. Horizontal shear acts along the grain and occurs, for instance, when a beam is subject to extreme bending. As the beam bends the fibres on the concave side are in compression, whilst the fibres on the convex part are in tension. Between the areas of compression and tension is a neutral area which is in horizontal shear. Under extreme conditions the fibres in this area can part, causing shear failure. Vertical shear acts across the grain and causes failure by tearing the timber in half. An example of a member that is subject to vertical shear is a jettied beam, which could break in half at the point where the lower wall supports it, if the jettied upper wall applied too much load on it. Vertical shear is unlikely because shearing resistance is much greater across the grain than parallel to it.

/\ **Bending** – Forces acting on a beam (or other member) produce a turning effect called a bending moment. This gives rise to compression in one side of the member and tension in other side. The bending moment can vary at points along the beam, and wherever the applied bending moment exceeds the moment of resistance for the member it will fail. Failure would be by crushing and/or tension in a part of the member cross section.

Truss Design

Trusses fall into two categories: those with tie beams, called 'closed trusses', and those without or with interrupted tie beams, called 'open trusses'. The principal purpose of a truss is to carry the roof loads from the purlins and to stop the walls of the building spreading outwards. This makes them a key structural requirement of any frame. They are, however, equally important as part of the overall design, as they help to define the internal structure of the building as well as offering the most interesting and beautiful pieces of joinery within the frame. In a standard conventional construction, trusses form part of the roof space only; however, in a structural frame they form a completely integrated part of the cross frame.

Most oak-framed buildings will contain a mixture of different truss types. The more elaborate trusses should be reserved for the open living areas, where they are going to make greatest visual impact. Practical trusses are best used to divide rooms or form gables. There seems little point in designing a frame where every truss is of an elaborate arch-brace design, if in most situations the truss will be partly hidden by the division of internal walls. In such a situation, it is much better to confine the elaborate truss to the open hall area of the main living space, where it can be appreciated and seen from all angles, and use simpler tie-beam-type trusses elsewhere. The following section outlines the most commonly used trusses, although it is not an exhaustive catalogue of every possible variation. With experience, it is possible to combine different elements of one type of truss with another, creating individual trusses to suit any particular situation.

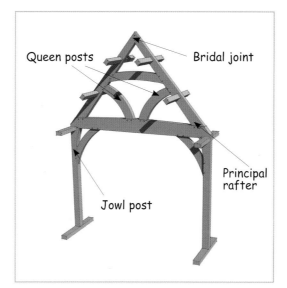

Queen posts · Bridal joint · Principal rafter · Jowl post

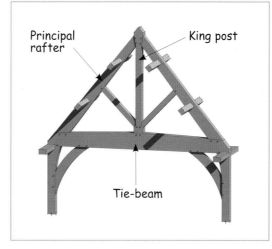

Principal rafter · King post · Tie-beam

Closed trusses – those with tie beams – can be split into two broad categories:

∧ **Principal rafter and tie-beam trusses** – This type of truss has many forms, and has been in common use since the fourteenth century. It basically consists of a large pair of principal rafters fixed with a bridal joint at the top and tenoned into a large tie beam at the bottom. The principal rafters support purlins and sometimes a ridge, which are either trenched over or jointed into them. To stop deflection in the principal rafters, by the roof loading, queen posts (or queen struts) are added and also sometimes a collar. These members are in compression, as they carry the load from the purlins down into the tie beam. These trusses may look simple but they can be used in a multitude of ways, because the internal framing can be arranged to suit many different applications. For example, a gable truss may be framed with vertical studs to take direct glazing or two queen posts and a collar to surround a window. Further refinements can also be added by making the tie beams curve or taper upwards.

∧ **King-post trusses –** These became very common during the eighteenth and nineteenth centuries and are still widely used today. Many people view them as the epitome of a traditional truss but in fact they were not widely used until the mid-seventeenth century. The main feature is the central king post, which is jointed into the tie-beam at the bottom and the principal rafters at the top, where it usually supports the ridge. King-post trusses work very well structurally and can span quite large distances due to the king posts acting like a keystone in an arch. A large amount of tensile force is produced within the king post by the upward thrust of the principal rafters. This tensile force on the king post helps prevent the tie beam from sagging and produces a tension joint. This needs careful designing at the joinery stage.

Above **Closed trusses, supporting purlins and wind braces in the roof**

Middle **Principal rafter and tie-beam trusses take many forms. This one shows a pair of queen posts and a collar**

Left **The king post in this truss acts like a keystone in an arch. The upward force on the king posts helps stop the tie beam from sagging but also produces a tension joint between them. This joint needs careful designing**

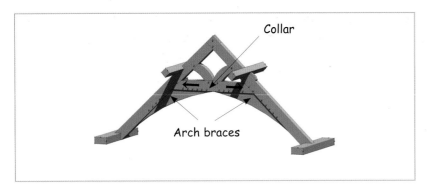

Collar

Arch braces

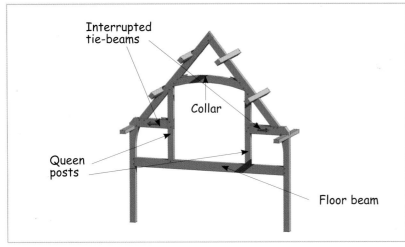

Interrupted tie-beams

Collar

Queen posts

Floor beam

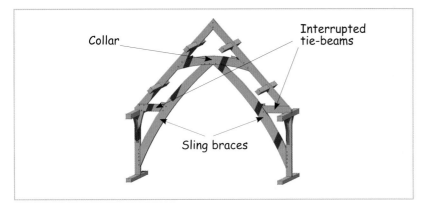

Collar

Interrupted tie-beams

Sling braces

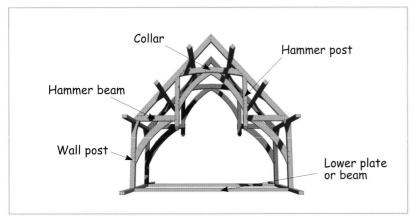

Collar

Hammer post

Hammer beam

Wall post

Lower plate or beam

Open trusses – those without tie beams or with interrupted tie beams – have four main variants:

⋀ **Arch-brace trusses –** These are usually employed to produce the vaulted-ceiling effect so common in our churches. Their basic form is a collared truss with two arch braces jointed beneath the collar. Because they do not have a tie beam to resist the horizontal force at the wallplate, the joints, which are in tension, have to be extremely well made to stop the truss spreading and therefore pushing the walls outwards. This is achieved by using dovetailed tenons or by inserting a discreet stainless-steel tie. Arch-brace trusses are usually employed to add a striking effect to a frame, so would commonly be used above the main living area to give a sense of height and drama. They are also useful when the tie beam on a normal closed truss would be too low to allow access from one bay to another.

⋀ **Interrupted tie-beam truss –** Quite often oak-framed houses are made only a storey and a half high, rather than two storeys, particularly if height is a planning issue. If this is the case a normal tie beam would fall below head height and impede access between bays on the upper floor. To overcome this a section of the tie beam is cut out, leaving the two ends (the interrupted tie beams) remaining and jointed into the principal rafters. To oppose the horizontal thrust at wallplate level, a queen post is jointed into the principal rafter and continued down to the floor beam below. The interrupted tie-beams are now dovetail-tenoned into the queen post to stop the truss from spreading, and the load is transferred through the queen post into the floor beam (or drop tie beam as it is also known). Because this is a tension truss, it is also a good idea to use a collar with dovetailed tenons to prevent any spreading.

⋀ **Sling-brace truss –** This is similar to an interrupted tie-beam truss, the main difference being that the vertical queen posts are replaced by angled sling braces, which are usually curved. Because the sling brace is angled, it doesn't have to be jointed into a floor beam and can be fixed into the main wall post of the cross frame. This is particularly useful because it allows you to use this type of truss on a single-storey building, where no floor beam is present. Sling-brace trusses work very well structurally and they also look excellent.

⋀ **Hammer-beam truss –** Hammer-beam trusses are open-type trusses designed for spanning large distances, often seen in the naves of churches, with a fantastic example in Westminster Hall, London, UK. The hammer beams are effectively formed by removing the middle part of a tie beam, which creates a large horizontal thrust on the frame. To counteract this thrust the wall post becomes part of the truss, by jointing a large brace between it and the hammer beam. Above the hammer beam a hammer post rises to the principal rafter, and it is also braced to a collar, so the loads can be effectively distributed to the ground. In the absence of a massive wall to oppose the force created at the foot of the wall post, it is essential that the posts are tied together with a lower plate or beam. The hammer beam should be two thirds the length of the wall post and the lower hammer brace should join the post at no higher than one third of its height and joint into the hammer beam directly under the hammer post. Both the lower and upper braces should be set at the same angle as the roof, as they essentially become a lower set of rafters efficiently transferring the loads.

Crown-Post Roofs

Crown-post roofs, unlike the previous trusses, are made up from common rafters with a collar jointed into them to stop deflection. The collar is supported by a crown plate (or collar purlin) which runs the length of the roof. This in turn is supported at intervals by a crown post jointed into a tie beam. This ancient form of roofing pre-dates purlin and truss designs. The crown post is braced to the crown plate and collar, to stop the whole structure collapsing like a pack of cards. Crown posts have traditionally been shaped for added decoration.

Far left top **An arch-brace truss**

Far left **An interrupted tie-beam truss**

Far left lower **A sling-brace truss**

Far left bottom **A hammer-beam truss**

Right **An example of a crown-post roof**

Below **The shaped crown posts in this roof support a crown plate, which in turn supports the rafters. Braces are added to stiffen up the roof**

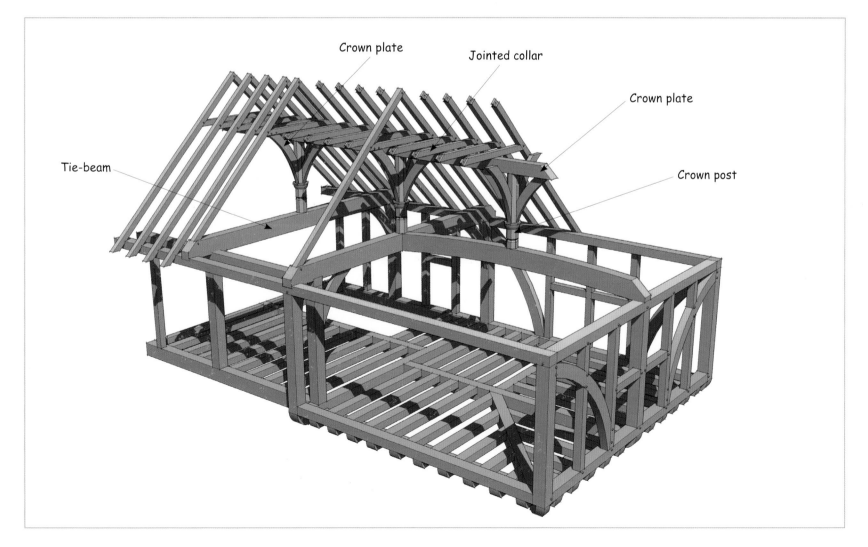

Cruck Design

We have already met crucks in the section about cross frames. Because of their construction, they naturally form part of both the walls and roof. Cruck frames can either be of the closed-truss type, if they have a tie beam at wallplate level, or of the open-truss type if they do not, in which case they normally have a collar at a higher level. Surviving cruck frames date back to the twelfth century, and there are many examples throughout Britain. Because their design varies so widely, the following list outlines some of the most common forms.

Some of the most common forms of cruck design:

∧ **Full cruck** – The blades of a full cruck reach right from the ground to the ridge of the frame, and are usually tapered from a wide bottom to a narrow top. They carry the entire load from the roof and walls into the base of the frame. How the blades are framed depends on the angle of the roof and how curved the blades are. As you can imagine, they are very large pieces of timber, quite difficult to come by, and if they are not sufficiently curved the wallplates and purlins have to be picked up by additional members, jointed into them.

∧ **Upper cruck** – Similar to a full cruck, except instead of rising from the ground, they rise from a floor beam or tie beam. Unlike the full cruck, they do not carry the entire load from the walls and roof, but act in a similar way to a sling-brace truss, transferring the loads into the floor beam and walls of the frame.

∧ **Raised cruck** – Here the blades rise from a solid wall, usually at least halfway up. Many of the great medieval tithe barns were built with raised crucks. A wonderful example of this type of structure can be seen at Winterbourne Barn, Gloucestershire, UK.

∧ **Base cruck** – A base cruck, like a full cruck, rises from the ground, but instead of continuing up to the ridge, it is jointed into a collar beam. This beam forms the lower member of a truss and the roof continues upwards with conventional framing. These frames are very useful either where wide spans need to be achieved or where long cruck blades are unobtainable.

∧ **Jointed or Somerset cruck** – This type of cruck employs two pieces of timber to make up a blade. The lower one springs from the ground and continues to above wallplate height, where it curves and forms the roof angle. A straight piece of wood is then jointed into it which continues up to the ridge. Many examples of this type of cruck have been found in the south-west of England, hence the name Somerset cruck.

Right **Raised crucks in Winterbourne Barn, Gloucestershire, UK**

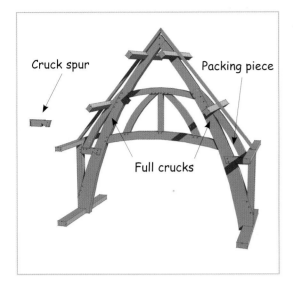

Cruck spur

Packing piece

Full crucks

Upper crucks

Wall post

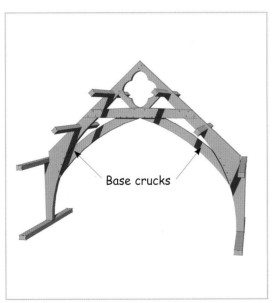

Base crucks

Above **A sling-brace frame**

Top left **A full cruck**

Left **An upper cruck**

Bottom left **A base cruck**

Below **In this gable frame the purlin is positioned so that the sarking on top of the rafters closes the roof to the inside of the building**

Purlins

Purlins are horizontal beams, which are set at right angles to the slope of the roof, and are generally attached to the principal rafters. Purlins carry the load from the roof into the cross frames, so their design has to be carefully considered. In order that they do not become too large in cross section, bay sizes should be restricted to around 13ft (4m). The number of purlins required on each side of the roof depends on the type and size of common rafter being used, and the distance between the ridge and the wallplate. Common rafters will only span so far without being supported. Also, as the pitch of a roof decreases, more load is transmitted onto the rafters, so the purlins have to be spaced closer together.

The position where the purlin is fixed into the principal rafter will determine where the top face of the common rafter lies. This will either be in line or above the top face of the principal rafter. If it is in line, the final roof covering will attach to both the principal and common rafters. Although harder to frame, this is advantageous on a gable truss because effectively the outside of the building can be closed off from the inside by the roof covering. Different types of purlin layout are listed on the following pages:

Trenched – A continuous run of purlins which are halved and notched over the back of the principal rafters. The most common form of connection is by the use of an interlocking dovetail, which helps stop the roof trusses from spreading. They can also be joined by a variety of scarf joints, but whichever method is used it is important to leave as much timber in the principal rafter as possible whilst still supporting the purlin. If the top surface of the purlin projects higher than the top surface of the principal rafter, it is a good idea to attach a cleat, to stop any rotation.

Clasped – As the name implies, the purlin is clasped between the underside of the principal rafter and a collar or strut. The principal rafters are usually smaller than those used with trenched or jointed purlins, because they are not taking as much load. The top surfaces of the principal and common rafters align, and frequently the principal diminishes in size to match the common rafter above the purlin joint.

Tenoned – There are a variety of ways of tenoning-in a purlin and, like the trenched purlin, the principal rafter has to be large to accommodate not only the load but also the purlin joint. The principal rafter's top surface aligns with the top of the common rafters (as in the clasped system). The purlins are usually jointed into the principal rafter with a housed tusk tenon, and are either jointed to form a continuous line or staggered at the principal. The purlins are staggered so as not to weaken the principals by taking too much timber out in a single location. Another very effective way of jointing-in purlins is to use a spline, which acts like a fish tenon by connecting two purlins on either side of a principal rafter. This is a strong joint and very safe when erecting a frame, as the principal rafters only need to be spread a short distance to engage the purlins. With normal tenoned purlins it is sometimes nearly impossible to spread the frames far enough apart to engage the joint.

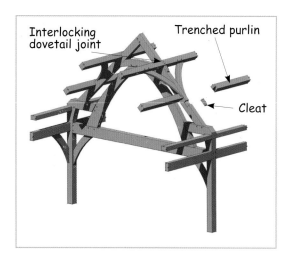

Interlocking dovetail joint · Trenched purlin · Cleat

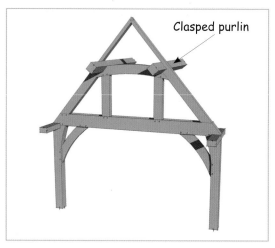

Clasped purlin

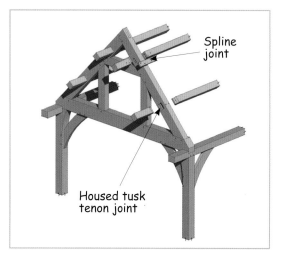

Spline joint · Housed tusk tenon joint

Above **Clasped purlins**

Top left **Trenched purlins. Purlins notch over the back of the principal rafter. A cleat is attached to the principal rafter to help stop any rotation in the purlin**

Left **Tenoned purlins. The purlins can be connected with either a housed-tusk tenon joint or splines**

Below **Both trenched purlins and jointed purlins are visible in this picture. Splines can be seen protruding from the gable truss awaiting the purlin ends**

At least one run of purlins on each side of the roof should be connected to a diagonal brace called a windbrace. These are jointed into the purlins and principal rafters and should be placed in pairs to stop the roof trusses from racking.

Hips and Valleys

When the slopes of two roofs meet, they form either an external or an internal angle. To connect the rafters an inclined timber is required which usually rises from the wallplate to the ridge. On the external angle this timber is called a hip rafter. It is capable of exerting a great deal of outward force on the corner of the wallplates, which can effectively be resisted by the introduction of a dragon beam and cross tie. Valley rafters form the opposing geometry to hip rafters and connect the roof planes on the internal angle, such as when two buildings join perpendicularly to one another. The framing of hips and valleys can be very complex, especially if the two connecting roof planes are of differing pitches to one another. If it is done properly it makes for a wonderful geometric pattern in the resulting roof space.

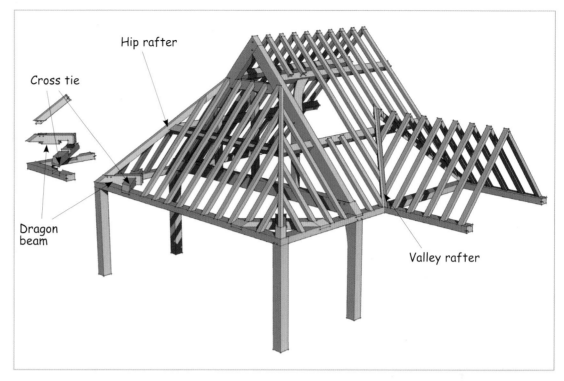

Hip rafter

Cross tie

Dragon beam

Valley rafter

Above **Dragon beams and cross ties help to resist the force exerted by the hip on the wallplates**

Right **A hip rafter**

Bottom right **A valley rafter**

Below **Curved wind braces like this one are connected to the principal rafter and the purlin to help stop the roof from racking**

Walls and Floors

As we have seen, purlins connect the cross frames in the roof plane. In the wall plane the wallplates and soleplates take on this function. These members should be wider than they are deep, and their width corresponds to the primary framing members in the wall. Typically if the main posts have a section size of 8x8in (200x200mm) then the wall/soleplate should be 8in (200mm) wide. As they usually run the whole length of the building, several pieces will need to be joined together to produce a continuous length, and this is achieved through the use of a scarf joint. There are many different forms of scarf joint, some of which are stronger than others, and their design depends on the type of loading they are likely to experience. The scarf joints should be placed near a main post, as this will help support the joint, and orientated so the frame can be erected in the correct assembly sequence.

Jowl Posts

A primary component that is common to both the cross frame and the wall frame is the jowl post. It is a primary structural post with a flared upper section where it joins into the wallplate and tie beam. The joint it makes with the wallplate and the tie beam is perhaps the most significant in the history of British-style timber framing and has been used extensively since the thirteenth century. The flared jowl post, being wider than the wallplate, is tenoned into the tie beam in the cross-frame plane, and also tenoned into the wallplate in the wall plane. The tie beam is also dovetailed over the wallplate, which helps tie the walls to the cross frames. The jowl post is traditionally cut from the base of an oak where the wood widens out, and is used upside down. The grain in the jowl naturally follows the curve of the trunk and produces the strongest possible timber. This is why jowls are also known as 'rootstock'.

Studs

Secondary timbers in wall frames include studs (which are vertical members) and transoms (horizontal members). These are smaller in section than the main posts and are used to define openings for windows and doors, or to provide a framework to attach the external cladding or infill panels. The term 'close studding' comes from the late-medieval practice of applying a large number of closely spaced studs to the outside of a building.

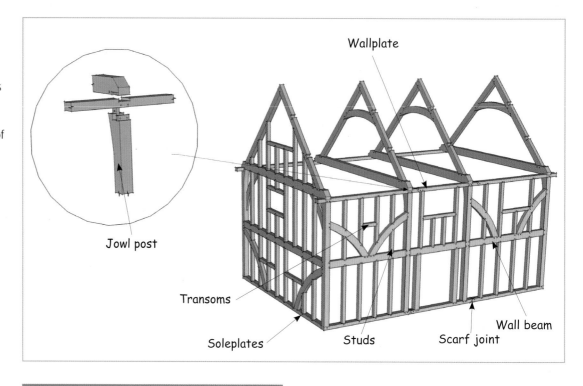

Above **Jowl posts are part of both the wall frames and cross frames. They are connected to the wallplate and tie beam by a three-way tying joint**

Below **Studs are smaller secondary timbers which help to define window openings or provide a framework for infill panels, such as in this 'close studded' house**

Wall Beams

If a wall frame is higher than one storey, then a wall beam (or edge beam) will need to be added to frame the edge of the floor. As this will be jointed into the main post at the same height as the floor beam, care needs to be taken that not too much timber is removed from the post whilst cutting the joints. Diagonal braces are also essential in walls to counteract the effects of racking on the frame; these can be straight or curved and if possible should placed in pairs. There is large scope for being artistic with braces, as the size and shape can be varied greatly, especially if curved timber is used. The depth of the braces needs to be considered if they pass behind the studs. For instance, if the wallplate is 8in (200mm) wide, and the stud is 5x5in (125x125mm), then the brace needs to be 3in (75mm) deep so that it doesn't overlap when jointed into the wallplate and main post. An alternative method would be to use a much larger brace, and joint the stud into it, such as in English Wealdon-style houses.

Below **Braces help to stop the wall frames from racking. Large braces like these are associated with Wealden-style houses such as Alfriston Clergy House, East Sussex, UK, a National Trust property that is open to the public**

Below right **Joists normally run between the cross frames in each bay. Unlike their softwood counterparts they tend to be laid flat, rather than on edge**

Joists

Most modern oak-framed buildings sit on solid foundations, and the final ground-floor surface is formed from that. In the upper-floor level, though, a system of joists are laid to span either between the cross frames (the most common method) or between the wall frames. Unlike modern softwood joists, oak joists are laid flat and are typically 6x5in (150x125mm) in section, but their size ultimately depends on the span between supports. As with purlins, once the span becomes greater than 13ft (4m), joist sections become excessively large, because of both the live and dead loads acting on floors. Normally this is not a problem when the joists are fixed between the cross frames, as the bay sizes should be restricted, to keep the purlins to a sensible section. When the joists span between the wall frames (in other words in the direction of the cross frames), then a bridging beam may need to be introduced to lessen the span. These beams tend to be large in section and need substantial tusk-tenon joints to secure them to the floor beams in the cross frames. Floor joists tend to be spaced between 16in (400mm) and 24in (600mm) apart, depending on their size and on the floor covering, and are jointed into the floor beams with a smaller version of the tusk-tenon joint, a simple full-depth housing, or a dovetail housing.

Why Oak Rafters and Joists are Laid Flat

Oak rafters and joists are generally different in section to their softwood counterparts. Typically they tend to be wider than they are deep. This traditional detail came about because they were commonly made out of small oaks (thinnings). Each tree would be split down the middle, yielding two rafters or joists. If the tree was bent, it was split so that the halved face was as flat as possible. This would mean that even though the rafter or joist was curved in one direction, it was flat in the plane of the roof or floor. When a halved log is further converted into a beam, the section produced will be rectangular, with the widest face on the heart and opposite side.

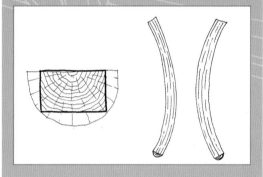

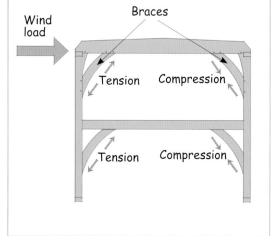

Jetties

Floor and wall design are combined in jettied buildings, where the upper floor and walls project further than the wall below. This type of construction was common in many framed houses built in towns as well as in high-status country dwellings. The amount a frame can 'jetty out' on each floor varies between 12in (300mm) and 24in (600mm) and depends on the size of joist being used. The joists extend perpendicularly from the wall frame and are supported by a jetty plate below in the lower wall, and have wall studs jointed into it to stop any sagging. The projecting joists in turn support the jetty bressumer plate which allows the continuation of the wall framing in the upper floor. If a building is jettied on more than one side then a diagonal floor beam called a dragon beam needs to be introduced, at the corner of the building, so the joists on both sides can project out perpendicularly to the wall frames.

Braces and Frame Stability

Stability is a fundamental part of structural design that influences the overall form of the structure more than any other aspect. Braces are an essential stabilizing component and the structure would simply collapse without them. They also play an important aesthetic role. Braces in British-style oak frames tend to be cut from curved pieces of timber, to follow the vernacular style of frame construction, whereas European- and American-style frames tend to use straight wood. This probably evolved because of the need of medieval carpenters to use lower grades of oak, often hedgerow material, as the best-quality timber was often saved for shipbuilding. Once selected, however, they would use the natural curves to make their buildings as decorative as possible.

The structural role of braces is to keep the frame square in the face of wind loading and other destabilizing conditions. Without them the frame would experience an effect called 'racking', where each beam or post would push its connecting member until everything fell down like dominoes.

Above left **The dragon beam and joists extend outside the building and support the upper walls in this jettied merchant's house (Leigh-Pemberton House, Lincoln, UK)**

Top right **The diagonal dragon beam enables the corner joists to project perpendicularly to the wall frames**

Above **Braces should be placed in pairs wherever possible, so one side is always acting under compression if the frame is under load. The brace which is under tension is trying to pull apart, whereas the brace which is under compression will only fail if the timber itself is crushed**

Opposite page **Polygon frames, such as this octagonal roof, make interesting spaces**

For braces to work effectively, they need to be distributed throughout the frame, so that they are working in every plane. There is no point bracing the walls and cross frames, but not the roof. It is equally important that they should be paired or mirrored on each beam wherever possible. This is not about symmetry (although they tend to look better if they are symmetrical) but because when the frame tries to rack over, one of the braces is placed in compression and the other in tension. The brace in tension is trying to pull apart, and its strength relies solely on the mortise-and-tenon joint with its connecting members. Whereas its opposite brace, which is in compression, will only fail if the brace is crushed.

Polygons

Most buildings have a rectangular footprint or are made up of a series of rectangles joined together, but it is possible to build houses that don't conform to this rule and are made up of more complex geometry. When designing multi-faceted frames the same rules apply for spanning distances, and care should be taken to avoid internal spans greater than 20ft (6m), otherwise the section size of the timber becomes unworkably large. The largest spans are likely to occur across the centre of the polygon in the floor plane. Taking a diagonal from one corner through the centre to its opposite point, the polygon should be scaled so this distance does not become too large.

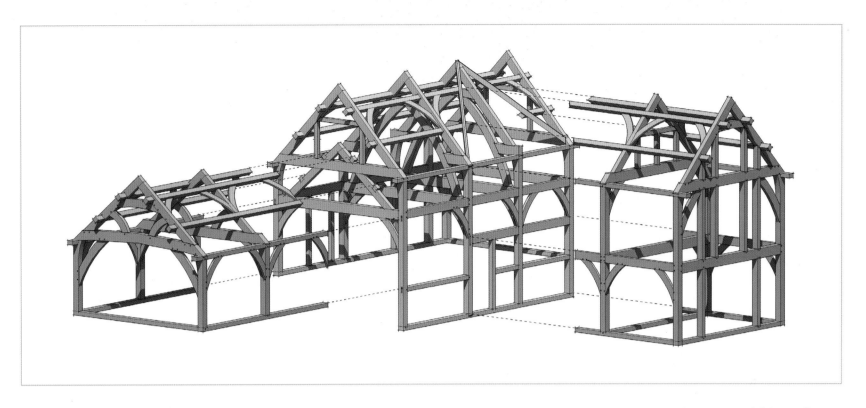

The Complete Frame

The previous sections in this chapter have outlined the various components that make up a complete structural frame, and one can see that frame design is fundamentally based on a grid system defined by walls and cross frames. At first glance this may seem to be a limitation but, believe me, there are an almost infinite number of combinations that may be used to connect the structural members within a frame. In its simplest form a house design is based on a rectangular floor layout, but even this basic shape can offer some interesting challenges to the frame designer. The bay sizes are able to be varied in order to accommodate the different living spaces within the building effectively, the walls can be jettied, and, with a large span the frame can be aisled. Add into the equation the different designs of truss that can be used, the joist layout or even the height of the walls and you can see that even with a straightforward layout there are many frame combinations available.

Connecting Frames

Oak-framed houses are ideally suited to the addition of cross wings and extensions, but design of these needs to be married with the principal structure. The best position for joining a perpendicular wing on to a main frame is at the primary framing members, so that the wing lines up with one bay or two bays on the connecting frame. This makes the construction

Above **When separate frames connect it is best to line up the primary timbers. This makes the valley rafters between the two frames symmetrical over one or two bays**

Far left **An example of a house based on a hexagonal grid**

Below **An example of a completed frame**

of the frame easier as the connecting timbers already have primary framing members to joint into. The junction between the two frames will produce a valley, and it is preferable to form this symmetrically within a bay – or at a cross frame, if it stretches over two bays – so the valley doesn't cut across a principal rafter in the adjacent frame.

If you plan to build on a sloping site, the foundations of the building will more than likely be on several different levels. The frame needs to be designed to accommodate the change in level in a way that won't compromise the integrity of its structure or design. The best way to achieve this is, again, to use a connecting frame that lines up with the corresponding bays in the main frame. In this situation the soleplates, and quite possibly the wallplates, will join at different levels. The foundation needs to be designed so that the common posts between the two frames extend to the lower level, allowing the soleplate on the higher level to be simply jointed into these posts.

Add-ons and Extras

Once the basic structural layout has been designed and all the principal framing members are in their place, the non-structural timbers can be added. It is very important to keep a clear distinction between what is structural and what is purely decoration. This seems to cause a lot of confusion, and many times I have been asked to remove strategic posts or braces by clients or designers once the plan has been finalized! These requests are usually connected with architectural changes to the house design and illustrate why it is important to understand the fundamentals of structural framing. The non-structural timbers on the other hand, have free range to roam and can be placed to help reinforce the architectural design.

A few ideas for non-structural timbers:

∧ Direct glazing studs added into wall frames and gable trusses

∧ Framing studs for windows and doors

∧ Dormers in the roof

∧ Internal partition walls

∧ Balconies and walkways

∧ Porches

∧ Handrails

∧ Decorative braces

Style

I haven't mentioned a great deal regarding the various styles of framed buildings, primarily because I prefer not to pigeon-hole them into different house types but let them stand on their own design and merits. But unfortunately (or maybe fortunately) not everybody shares my view. What follows are the most common classifications.

Below **A barn-style oak frame**

Different styles of framed buildings:

∧ **Traditional** – 'Olde worlde' appearance with lots of timber in evidence. This is most people's idea of a timber-framed house – the classic design seen in so many British towns and villages. External walls have exposed studs and transoms, and between these rendered panels are fixed. Jetties are also a feature, and so too are large curved wall braces.

∧ **Barn style** – The main difference between this and the traditional style is that barns are normally clad externally, so there is less need to have many studs on which to attach infill panels. Ironically though, they do tend to incorporate large areas of direct glazing, which is fixed to studs. Internal spaces tend to be left open as much as possible, with the framing in evidence everywhere.

∧ **Post and beam** – Normally this refers to a more American style of framing, with fewer secondary timbers and hardly any curved components. The timber is framed in a different way and Douglas fir and larch are more commonly used than oak.

Below **A post-and-beam frame built in Douglas fir**

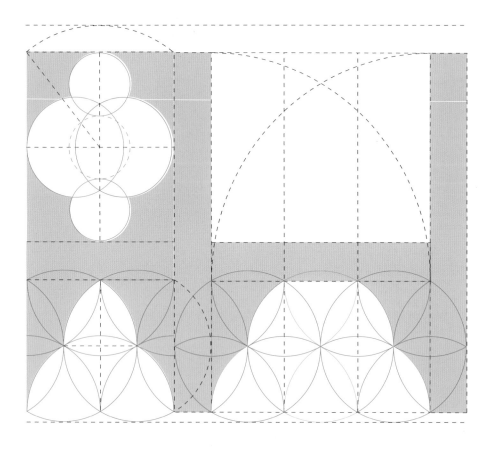

a seat cut (which brings it flush to the top corner of the wallplate) or a birdsmouth (which is cut to one third of the perpendicular depth of the rafter). The underside of the common rafter now gives the position within the roof of the purlin and, depending on how it is jointed with the principal rafter, the position of that as well. The tie beam is dovetail-notched a set distance over the wallplate and the principal rafter is jointed into it. Sometimes the positions of the principal rafters will vary within a frame if a mixture of purlin jointing details are used, for instance trenched within the middle of the frame and jointed at the gables.

Building Regulations and Engineers' Calculations

Once initial permission has been granted, another set of drawings needs to be submitted to the local authority for approval. These contain a lot more information than the planning drawings, and detail every stage of construction with accompanying written notes. All this information has to comply with the latest building regulations, and the constantly changing standards, such as higher levels of insulation for instance, have to be reflected in every new design to gain approval. Of course this has a cost, which is usually split into two parts: firstly for approval of the drawings, and secondly for on-site inspections by the building inspector to check that the work is being carried out correctly.

The frame design is either incorporated fully within the building-regulation drawings or supplied as a supplement to them, as the structure obviously comes under close scrutiny by building control. As oak framing is not standard construction, they normally require proof by a qualified structural engineer that the frame is capable of doing its job properly when it comes to resisting the loads and forces applied to it. The process is quite straightforward but does require some pre-planning. The frame design is

Geometry

To design frames effectively, not only does the mathematical relationship between the individual members need to be understood, but also the scale and proportion of the building. Medieval timber framers were highly skilled in the use of geometry, and evidence of daisy wheels used to calculate the proportions of the frame has been found on some ancient timber structures. For any building to look well designed its proportions need to be correct, so that it looks comfortable to the eye. This is no easy task and it is where a well-trained architect or designer comes into his or her own.

Floor Heights

In architectural plans it is normal to measure floor heights in terms of the finished floor level (FFL). When designing a frame the floor construction needs to be taken into account to arrive at the correct height for the floor beams and joists. Different types of floor construction are covered in Chapter 8: Finishing. In most situations the soleplate will sit on the structural slab with insulation between it and the flooring surface slightly above. On the upper floor the

insulation and flooring will all be above the joists and floor beams. If the distance between the floor levels and the make-up of the floor is known, it is a simple calculation to arrive at the floor-beam height. If there is a floor within the roof trusses then the distance between the finished floor levels will also set out the wallplate height because the tie beam is normally dovetailed over the wallplate by a set amount.

Roofing Geometry

The roof is set out from the wallplate and as we have seen the height of it can be determined from an internal floor or perhaps from a planning restriction on the total height of the building. How much the roof rises from the wallplate is determined by the span of the building and the pitch (angle) of the roof. The greater the pitch, the steeper the roof will be and therefore the higher it will rise for a given span. The best way to lay out the principal framing members is to start from the outside and work in, the various positions of each being determined by which carpentry detail is used. To begin with, the common rafters will fix to the wallplate with either

usually completed after the planning drawings, at which stage they can be incorporated with the building-regulation drawings and also sent to the structural engineer. The engineer will then produce a set of calculations proving the structure and checking the members for structural adequacy. If the timber sizes are too small or not connected in the right way, recommendations will be made which are sent back to the frame designer, who will then make the subsequent changes to the design. The calculations, frame design and building-regulation drawing can then be sent off to the building control department for approval, with the correct fee of course!

Apart from the statutory requirement to have the frame checked for building regulations, it is essential, unless your frame designer is a qualified structural engineer, to engage one. I cannot stress this too highly, as oak-framed buildings contain too many large, structural pieces of timber to mess about with. Once an engineer has checked the frame, his professional indemnity insurance will cover the structural design, so long as the workmanship and specifications are carried out correctly. This isn't always quite as simple as it seems, as engineers, just like other professionals, tend to specialize in different areas and many may have no experience of traditional framing. If you do experience problems finding one, try contacting the Carpenters Fellowship in the UK at www.carpentersfellowship.co.uk for a list of suitable engineers. In the USA the National Council of Structural Engineers Associations (NCSEA) is a good starting point at www.ncsea.com or the Timber Framers' Guild at www.tfguild.org

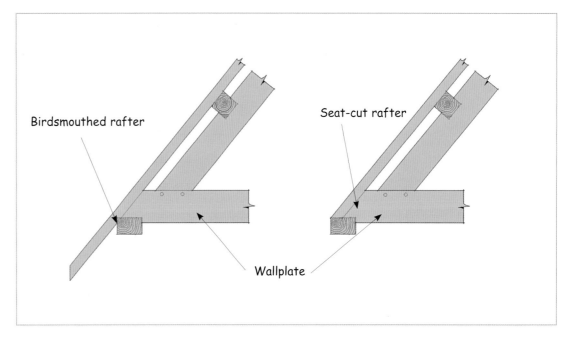

Birdsmouthed rafter

Seat-cut rafter

Wallplate

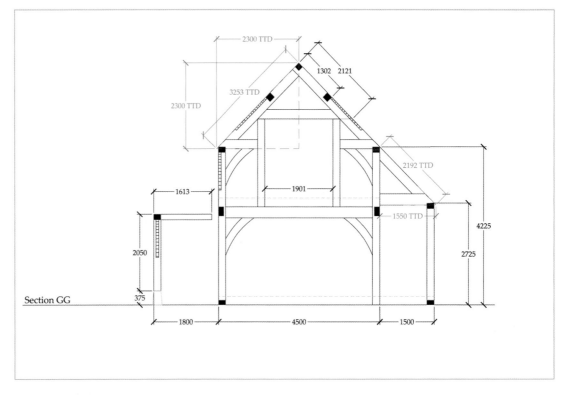

Section GG

Far left **Daisy wheels were often used by medieval timber framers to design their frames**

Left **An example of a Daisy Wheel design used in a building in the Forest of Dean**

Top right **Roofing geometry is set out from the wallplates. The diagram on the left shows the rafter birdmouthed over the wallplate, whilst the one on the right shows the rafter having a seat cut**

Bottom right **Once the design has been finalized, working frame drawings can be produced**

Frame Design

∧ **Grid** – Oak frames are arranged on a grid system, where the principal framing members are grouped together.

∧ **Cross frames** – Loads imposed on the building are transferred via horizontal beams into cross frames, which transmit them into the foundations.

∧ **Bays** – The gaps between cross frames are called bays, and their size can vary within a building, but ultimately they are restricted in length by the section size of the load-bearing horizontal timbers within them, such as purlins. It is best to keep this distance between 10ft and 13ft (3–4m).

∧ **Beam lengths** – Tie beams and floor beams tend to be the largest clear-spanning members within a frame, but once their length exceeds 20ft (6m) it is good practice to introduce a load-bearing post to reduce the effective span, otherwise their section size will become too large.

∧ **Braces** – To counteract the effects of lateral loading on the frame, such as wind, braces are introduced to stop it from racking (causing a domino effect). Braces should be placed in pairs wherever possible, and placed throughout the building in the cross frame, wall frame and roof planes.

∧ **Check** – Always have your frame design checked by a properly qualified structural engineer.

Right **The cross frames in this picture carry the roof loads via horizontal beams into the foundations**

Left **A contemporary post-and-beam style frame in oak**

The Minstrel Frame

The Minstrel Frame was built on the site of an old falling down bungalow on the edge of the South Downs in Sussex. The inspiration for its design came from the shape of the site, which it now sits in, and the clients who come from a family of traditional folk singers. The architect's concept was of an octagonal-type building with a minstrel gallery along the south side where the family could hold performances. The roof extends over the minstrel gallery to form a covered balcony. The south side of the building was also designed with a lot of glass to gain from passive solar heating whilst the overhanging roof shades the glass from over heating in the summer. The north side of the building was designed with minimal glazing to restrict the loss of heat.

Above **The posts in the octagonal section of the building are pre-shaped to make the framing easier**

Below **The south side of the building was designed to be full height and glazed in order to achieve both views and gain energy via passive solar heating. The overhanging roof protects the building from overheating in the summertime**

To translate the design into a structural frame an octagon grid was created, extending into two rectangles. The main living space is situated on the first floor within the octagonal end of the building. This had to be completely open, which made for a challenging roof design. In the end a type of crown post roof was used, with the crown plate being supported by two large posts that stretched from the roof to the foundations. The rafters were strengthened by a collar that rested on the crown plate.

The main posts of the octagon section were given a five-sided shape by using a portable bandsaw mill. This helps to tidy up the framing, allowing all the connecting beams to be jointed perpendicularly to the posts. When an octagonal frame is made in the workshop, each face of the octagon has to be made separately. The shaped posts have to be wedged so that they present one horizontal and one vertical face to the side that is currently being made.

The frame was finished on the outside with locally sourced sweet chestnut cladding. This was laid horizontally with its top and bottom edges angled, so each piece is slightly overlapped. They are not fitted tightly together so air can circulate behind the boards but the angled edges stops the wall behind from being visible.

Above **The roof is supported by a crown plate and two posts which reach to the structural foundation**

Right **The framing drawings were formed from an octagonal grid. The drawing shows the main floor beams and joist layout at first floor level**

Far right **The rafters were constructed with a collar sitting on the crown plate. The collar helps to stiffen the rafter and should be placed low enough to stop any deflection**

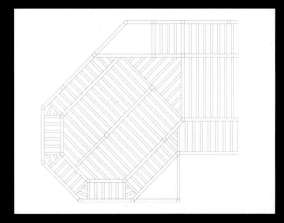

Left **Oak has many useful qualities, which has made it Britain's most important broadleaf, and this in turn has led to the success of British-style oak-framed buildings**

Chapter 4

The Structural Qualities of Oak

Not only is the oak a beautiful tree but the timber it produces has many useful qualities. Oak came to Britain about 9,500 years ago after the last ice age, and during the following 3,500 years spread northwards to inhabit most of the island. The warming during the post-glacial period also encouraged animals to move northwards, which were in turn followed by hunter-gatherers. As the population increased, oak became an ever more useful resource because of its strength and durability. Indeed evidence dating back as far as 7,500BC has been found of oak being used as a building material. It also became useful in other industries and large tracts were felled to produce charcoal for use in smelting during the Bronze Age.

Above **Widely spaced oaks in a well managed woodland**

A History of Oak in Britain

The many useful qualities of oak have made it Britain's most important broadleaf tree in forestry terms, and this in turn has led to it becoming culturally significant. Thousands of years ago Druids considered oak trees to be sacred, especially if they had mistletoe in them, and they would perform many ceremonies under their spreading boughs. As forests were cleared to produce land for agriculture and fuel for charcoal, single oak trees were often left to grow to a great age. These trees became important meeting places, and were often given names, many of which still survive today.

The importance of oak to the ancient economy cannot be overstated. Forests were managed to produce timber, and because it coppiced so well and the timber was so useful oak became more dominant than any other species. Another major factor in the success of oak was how easily it could be processed into a finished product. One of the best ways to convert a log is to split it along its length into sections, using wedges, rather than cutting or sawing it – and oak will cleave better than most timbers because of the conspicuous medullary rays, which radiate out from its heart. Cleft wood is very strong and durable because the split that is formed follows the grain of the wood and the fibres are parted rather than cut. Oak that was processed this way was used to produce durable fence posts, beams for buildings and ships, shingles for roofing, laths and wattles, wheel spokes, chairs and poles of every description. The bark of the tree was extensively used for tanning (curing leather), and acorns from large, well-spaced trees were used as a food source for animals. This custom, called 'pannage', is still practised in the New Forest in Hampshire, UK, today, where pigs are released into the woods, in autumn.

Left **Oak trees can survive for many centuries. Ancient oaks have often been used as meeting places and many have even been given their own name**

Above **Oak being cleft for lathes. The ability of oak to cleave easily, combined with its durability, has made it an important commodity**

Far right **Felled logs waiting to be transported to the sawmill**

During the Roman period of occupation 2,000 years ago, many of Britain's forests were decimated for the production of charcoal, which was used in the smelting of ores and in glass-making. Oak was also used for shipbuilding, construction and for heating buildings and baths. Following the Roman departure, the climate in Britain cooled, causing crops to fail, and famine and plague ensued. The chaos of the times gave the forests a chance to revive naturally until they were being heavily exploited again during the medieval period. Demand was again reduced with the advent of the Black Death, but afterwards the forests were decimated again for shipbuilding and the production of charcoal for smelting. By the eighteenth century much of the native forest had disappeared and the Royal Navy first turned to imported timber for the construction of its ships, and then to iron. Many forests were turned into agricultural land and managing oak forests became uneconomical, so by the beginning of the twentieth century only about 5 per cent of Britain was covered with woodland. The last century has seen a dramatic improvement in forest cover, even though massive amounts of timber were used during both world wars. Now it stands at nearly 12 per cent, of which the largest standing timber volume of any species is oak.

Coppicing

Cutting trees down to just above the ground, so that they can naturally regenerate. This will produce long poles rather than a thick trunk.

Oak and the Environment

Europe as a whole has seen its forests grow in size over the last 100 years, and much of this can be attributed mainly to improvements in forestry and silviculture. It is now estimated that the annual harvesting of timber in Europe is a quarter to a third lower than the net annual gain in timber volume. In Britain it is estimated that only half of the volume of hardwood that grows each year is harvested, so it would be possible to double the output without any net loss. This bodes well for the future because not only is timber a renewable resource, but our home-grown supply is actually increasing. This is also the case in the US where the hardwood stock has been rising by about 1.7 per cent every year for the last 30 years. In Britain the Forestry Authority, which is part of the Forestry Commission, sets national standards for the care and management of broadleaf woodlands and also administers grants to encourage further planting. Many woodlands now produce FSC (Forestry Stewardship Council) certified timber or PEFC (Pan-European Forest Certification) timber, which gives a chain of custody from the forest to the customer. In that way, you know that the timber you have bought has been produced in a sustainable way. Although these organizations have their critics, it is still a massive step in the right direction.

The Greenhouse Effect

Greenhouse gases increase warming in our atmosphere, which could lead to dire environmental consequences unless their progress is checked. The main greenhouse gas is carbon dioxide (CO_2), which is constantly increasing because of our use of fossil fuels. Trees, on the other hand, act as a 'carbon sink' and they absorb CO_2 during photosynthesis. This locks the CO_2 within the tree, until the tree dies and decays, or it is burnt as fuel. This does not increase the net CO_2 levels in the atmosphere because only the amount that was absorbed by the tree in the first place is released. Overall CO_2 emissions in the atmosphere can be reduced by replacing fossil fuels with carbon-neutral wood, but if felled timber is converted into something that has a long-term use (longer than the life of the original tree) then a significant amount of CO_2 is removed from the atmosphere. Oak frames are a perfect example of this, because the average life of an oak tree used in a frame is between 60 and 100 years, and yet if the building is looked after properly, it will last many hundreds of years.

Embodied Energy

Manufacturing green-oak frames requires less energy than other forms of general construction made out of alternative products because its embodied energy is so low. Embodied energy is a measure of how much energy is required to extract, manufacture, treat and transport the material to where it is going to be used. For example, the approximate embodied energy of locally produced green oak is 220kWh/m^2, whereas steel is 24,700kWh/m^2 and aluminium is 141,500kWh/m^2. Unlike other timber in construction, green oak does not need to be treated extensively with preservatives, which have been known to pollute soil and watercourses.

The Structure of Oak

To understand the structural properties of oak and how they affect the final design and manufacture of an oak-framed building it is important to know how the tree grows and what the characteristics of the material are. This is especially important with oak as it tends to be worked green, in other words unseasoned, and is therefore subject to movement and shrinkage as it dries out. There are many reasons for using green oak, rather than dried or seasoned oak, but the primary one is the rate at which timber dries. The carpenters' rule of thumb for the drying rate of oak is an inch (25mm) a year, but in reality the actual drying times can be a lot longer and depend on many factors. It is true that for the first couple of inches oak that is stacked properly will dry out at approximately an inch a year, but thicker pieces take proportionally longer to dry out. So while a section of oak 2in (50mm) thick might take only

two years to air-dry, a 4in (100mm) section could take five or six years. The section sizes used in an oak frame tend to be much larger than 4in (100mm) and can take many years to become dry, for instance a large tie beam could take as long as 20 to 30 years to dry out. This means that one could wait many years for the timber to be dry enough to use, even if the stockpile of correctly sized and dimensioned timber was available. There are cost implications as well, as the value of a piece of dry timber can be as much as four times that of an equivalent piece of green timber, due to the processing time involved.

Different Types of Oak

There are two native oaks in Britain (although they are not just confined to the British Isles), English oak or pedunculate oak (*Quercus robur*) and sessile oak or durmast oak (*Quercus petraea*). Both have

very similar structural properties, although in the past the Royal Navy believed that pedunculate oak was better than sessile oak for planking their warships. This was because the longer fibres found in pedunculate oak were better when it came to resisting cannon balls. It was a long time before anyone realized that the reason that pedunculate oak had longer fibres was that it was planted further apart than sessile oak, and therefore grew faster. Oak is classified as a hardwood (angiosperm) and is in most cases a broadleaved deciduous tree. Hardwood trees tend to be denser than softwood trees, although there are some obvious exceptions: the tropical balsa wood tree is a very soft hardwood! Oak is also deciduous, which most hardwoods are, but again there are exceptions such as holly, which is an evergreen.

The Anatomy of Oak

Trees have a very complicated structure that extracts water and minerals from the soil and carbon dioxide from the air to produce food, which helps the tree grow. The water is taken in by the roots of the oak tree and drawn up through the trunk and branches to the leaves, via a series of hollow straw-like cells known as vessels. The leaves absorb carbon dioxide during the daytime, which is combined with the water, and, using energy from the sun, make basic sugars in the form of carbohydrates. During this process, known as photosynthesis, oxygen is expelled and the sap carries the sugar back down the tree through the bast, just beneath the bark, to form new cells in the cambium layer. The miracle of this chemical conversion is made possible by the green chlorophyll present in the leaves.

Oak has two types of growth through one growing season. In the springtime it has a period of rapid growth called earlywood which produces large vessels, whilst during the summer the latewood growth creates much smaller vessels. The different sizes of vessel produced during the spring and summer cause the annual growth rings to look more pronounced, and timber of this sort is said to be 'ring-porous'.

Apart from vessels, the other main types of cells that make up oak timber are the fibres, which are long, thin vertical cells running parallel with the trunk, and the rays, which are horizontal cells radiating out from the centre of the tree. The fibres are tightly packed together and make up most of the woody tissue, their main function being to provide strength and support for the tree, whereas the ray cells (called medullary ray) transport and store food used by the tree. It is these rays that have proved so useful in the past because they offer a line of weakness along which the timber can be split. Because wood is made of different types of cells, it is said to be anisotropic, which means its structural properties are not uniform in each direction, unlike, for example, plastic.

Far left **Large-section oak can take many years to dry. Normally it is used green within a frame and dries *in situ*. Green oak is cheaper and easier to work than dried oak**

Left **Pedunculate oak (*Quercus robur*)**

Top right **Trees are nature's factories, producing sugars and oxygen whilst absorbing carbon dioxide**

Right **A section through an oak tree**

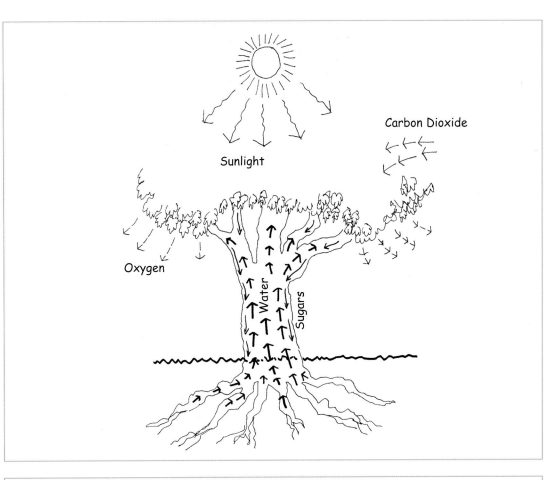

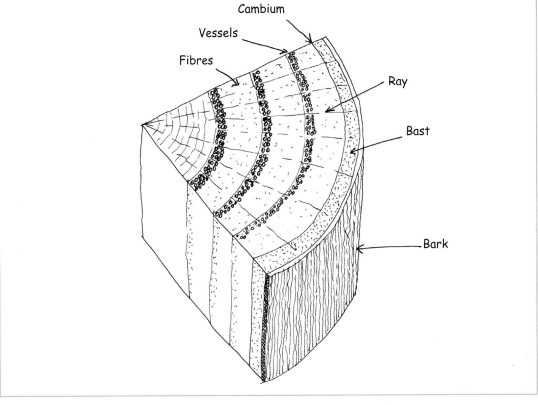

The outer area of the tree where the sap flows is called the sapwood. As the oak grows and lays down new rings each year, the inner cells nearer the centre undergo a chemical transformation. They stop contributing to the growth of the tree but they do continue to help support it. The inner or older cells build up deposits of tannin, a chemical that turns the timber a darker colour. This part of the tree is called the heartwood, which is much more durable than the sapwood, both in its resistance to rot and insect attack, because of the toxicity of the tannin within it. Most of the oak used for frames contains hardly any sapwood, at most some on the corners of large beams. Many medieval buildings, which used lower grades of oak, contained a higher proportion of it. This can be seen in old beams that have lost their rectangular shape and look almost rounded. The sapwood has been eaten away over the years, leaving the hard heartwood intact.

Cultivation

The best oak for structural use is grown either to have a long, straight trunk with no side branches until the crown is reached, or naturally curved timbers that are useful for cruck construction (see page 52). To obtain good-quality, straight oak it is important that it is grown correctly with proper forest management. Creating the correct environment is essential so the trees are forced to grow upwards quickly, producing few or no side shoots. This is done by inter-planting or under-planting the oak with nursery trees, such as Norwegian spruce or hornbeam, to create side shade and a secondary crop. As the trees grow larger both the nursery and oak trees should be thinned out, so the crowns of the oak get enough light. Eventually, the forest should contain large, widely spaced, fast-growing oak trees, with a dense under-storey of shade-loving plants to discourage any side shoots in the oak.

Slow-grown oak is very weak because its cell structure mainly comprises earlywood, which has a high level of large, light vessels within it. Fast-grown oak, on the other hand, has wide rings containing more latewood growth, which comprises more of the denser fibres and less of the light vessels. The denser a timber is, the greater is its strength. Slow-grown oak has thin growth rings which contain proportionally more earlywood, made up of large, light vessels. Fast-grown oak, on the other hand, has wide rings containing extra latewood growth, which comprises more of the dense fibres and fewer of the light vessels. This situation is reversed in softwood, where the fast-grown timber is weaker (less dense) than slow-grown wood.

Traditionally, curved timber was highly sought after and specially grown (mainly for the shipbuilding industry) by tying down the tops of young oak trees to a nearby tree, preferably to the south of it, so that it would naturally grow in that direction, rather than trying to straighten itself. Today most curved timber comes from wind-blown trees or trees that are grown on a slope. Initially the tree grows out at an angle and then naturally straightens itself to grow upwards. Some curved timber may be obtained from the branches of a large tree, but care needs to be taken that it does not contain any large knots. Curved timber often contains 'reaction' wood, which is produced because one side of the tree is in tension, and the other side is in compression. The effect of this is to produce an off-centre heart in the timber, with the rings tightly spaced on one side and widely spaced on the other. Reaction wood can shrink along its length, whereas in normal timber this movement would be negligible. The result of this shrinkage can be an increase in the curvature of the timber once it has been cut.

Far left **The outside, sapwood part of oak is more prone to insect attack than the heartwood. Although many old beams look eaten away, it is usually just the sapwood on the outside. Internally the timber can be as hard as iron and unaffected**

Left **Good-quality straight oak is grown by forcing it up and restricting side shoots. This is done by inter-planting it with other species of tree such as hornbeam or spruce. As the trees get larger the oaks and nursery trees need to be thinned out**

Right **Curved timber can be obtained from trees grown on a slope. The tree initially grows out before curving up to grow vertically**

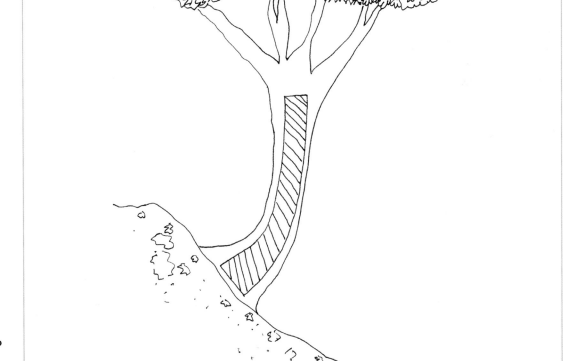

Conversion

Today most oak used in framed buildings is purchased from specialist sawmills, or extracted locally and converted into beams with the aid of portable bandsaws. To convert the logs into beams, they first have to be felled, trimmed of any side branches, and transported to where they are going to be processed, which is a job normally done during the wintertime. Winter felling is important for many reasons, listed right.

Far right **An oak log being hewn with a broad axe**

Below **A portable bandsaw is very useful for converting logs when it is difficult to transport them to a mill**

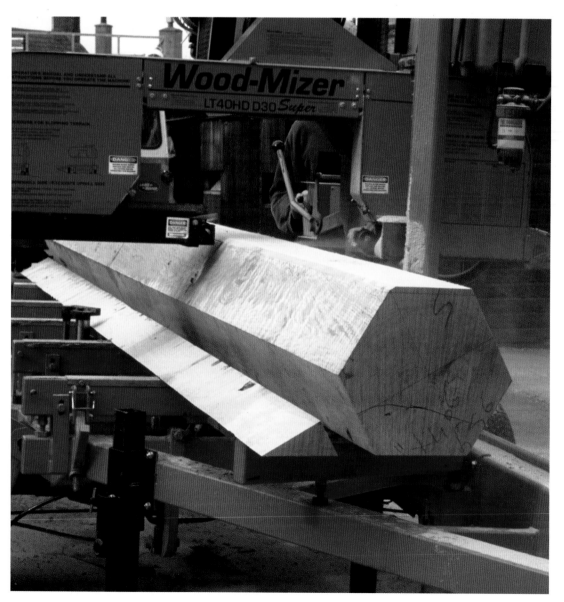

∧ **Insect resistance** – After the leaves have been shed there is no sap rising within the sapwood. As you can imagine, during the summer, while the leaves are out and photosynthesising merrily away, the sapwood is full of sugary starches that will remain within the timber after the tree is felled, and provide a hearty meal for any passing insects.

∧ **Lower moisture content** – The sap also contributes to the overall moisture content of the tree. The timber with a higher moisture content has a tendency to dry proportionally faster than that with a lower moisture content, and in turn causes more surface checking (splits).

∧ **Slower drying rate** – The temperature is colder during the wintertime, and this helps to slow down the rate of drying within the timber, which again makes it more stable and less prone to checking.

∧ **New sapling growth** – The leaves have fallen off the small branches, so that once they have been trimmed off the felled trees and left to rot on the forest floor, they do not completely cover the ground and restrict new saplings' growth.

Many countries around the world still practise felling trees by the moon cycle (no, I haven't been drinking 'moonshine'), including parts of Europe and South America; unfortunately it seems to have died out in Britain. Our ancestors believed that the best time to plant crops was on the waxing moon (just after the new moon) when the sap was rising in plants and trees, whereas the waning moon (just after the full moon) was the best time to fell trees and trim shrubs as the sap level was declining. Looking at it scientifically, a link between tide levels (which are affected by the moon cycle) and the level of sap within trees has been proved, so a tree that is cut during the waning period should have less sap and therefore be less prone to insect attack.

Boxed-heart

A boxed-heart is the result of squaring off a tree in order to leave a beam of heartwood.

Traditional Methods of Converting Logs

Traditionally, logs were converted into usable beams by one of three methods: hewing, sawing (pitsawing) and splitting. Traditional conversion is still practised by some framers today, whose clients want a totally hand-finished product.

Hewing

This common method uses axes to convert a round log into a boxed-heart, square or rectangular beam, and evidence of this process can be seen in old buildings whose timber has a rippled texture. This is often mistakenly said to have an 'adzed finish', when in fact adzes were never used to convert logs but were employed in shaping and finishing smaller timbers such as joists and floorboards. The usual way to hew a log is to secure it to the required working height, by using two large metal staples – called 'dogs' – to prevent it rocking. Then the required dimensions of the finished beam are drawn on each end of the log, using a level to keep the lines horizontal and plumb. These lines can then be joined together over the length of the logs by snapping a chalk line (the edge away from the dogs is done first). The first of two axes, called a felling axe, is now employed to remove the bulk of the waste timber up to the marked line, by scoring the edge using vertical angled cuts. If a lot of waste needs to be removed a series of vertical notches called joggles are cut to the line first, and then the waste wood between them is simply split off. The beam should now have a rough flat edge, which is dressed smooth with the aid of the second axe, called a broad axe. These axes resemble the ones used to cut off the heads of kings and queens during the Middle Ages, and are usually bevelled on both sides to create a razor-sharp edge, to slice down across the grain of the timber. It is this axe that gives the timber the so-called 'adzed look'. Once one edge has been completely finished the log can be rotated by a quarter turn, and the process continued until all the sides have been finished. It should be pointed out that this is a very skilled and dangerous job, and should only be attempted by trained professionals.

Adze

A tool that is used for shaping or smoothing timber. It has a thin, arched blade at right angles to the shaft.

Sawing

Timber has been sawn into beams since the Roman era, either by using a pitsaw, where the logs are rolled over a man-made hole with a supporting framework above it, or more commonly by placing the logs on large trestles. The advantage of the latter method was that the trestles could be moved to the heavy logs where they could be sawed *in situ*, rather than dragging the logs to the pit. The ripsaws used are large two-man affairs, where the man on top of the log – the 'top dog' – guides the saw along the marked line, whilst the poor bloke beneath the log – the 'underdog' – provides the muscle, and gets a faceful of sawdust by way of thanks. Hand-sawing in this way can be quite a fast method of conversion, and was still commonly in use up until the last century. Each stroke of the saw can cut through up to half an inch (13mm) of timber, leaving an inclined mark on the timber very similar to the marks left by modern bandsaws (although these tend to be vertical). Unlike hewing, the saw can convert the timber into halves, quarters or slabs.

Splitting

As mentioned earlier, oak splits very well, and halving a log is a much faster method than sawing; the resulting timber being preferable because none of the fibres in the oak are cut, as they would be if sawed. This produces a much stronger timber. The logs are split or cleft by banging in wedges from the end of the log first, and then as it opens up, down the sides. The split will naturally follow the heart of the timber, but in larger logs this can follow quite an unpredictable route. Also, the grain can have a tendency to spiral, so that when the log is finally halved it can have a pronounced twist and be unusable. For this reason only the smaller oak trees and poles are normally cleft to be finally converted into members such as rafters and joists. Logs should be cleft taking account of any curve in the timber, so that the two split faces are as flat as possible, and the curve is taken up on the side. Once the sides and bottom have been squared, these members can be laid flat in a floor or roof, with the widest dimension being in the width rather than the depth. This is the reason why medieval oak frames tend to have their rafters and joists laid flat, rather than on edge, as in modern construction, and most new oak frames follow the traditional convention because, I suppose, it just looks right.

Modern Conversion

Today most oak is converted into beams by means of a large bandsaw with a carriage rig. After the trees have been felled and extracted from the wood, the logs are transported to a sawmill. They are stored until such time as they are needed, and, depending on the size and shape of the log, this can be up to a few years (or even longer in the case of specialist curves and crucks). When the log is eventually converted, it is placed on the carriage which moves the log through the bandsaw blade, reducing it to the required section needed by the timber framer. This in itself is a very skilled job, as a lot of knowledge is needed to achieve the best and most economical use of the log to produce the many various sizes of timber needed in a whole oak frame.

Another popular method of modern conversion is to use portable bandsaw mills, which can be towed behind a vehicle to convert the logs where they are felled. These are particularly useful for small woodland owners and farmers who want to convert their own timber without having the hassle of transporting it to a sawmill. Unlike the large, fixed bandsaws, the logs tend to be placed on the bed of the carriage and the head of the saw moves along the log.

An average tree will produce about 3m² of timber (lumber), of which about half is converted into large beams. The rest is converted into smaller sections of timber, which is used, for example, in joinery or flooring. A standard three-bedroom oak-framed house will use approximately 270ft² (25m²) of oak in the frame, and if each tree produces about 16ft² (1.5m²) of beam-quality oak, then roughly 17 trees will be needed to construct the frame.

BOXED-HEART
Members converted this way include posts, beams and plates.

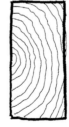

HALVED
Members converted this way include plates, principal rafters, tie beams, crucks and braces.

QUARTERED
Members converted this way include rafters and joists.

All photos left **Oak logs are split by banging in large wedges. Cleft wood is stronger than sawn wood because fibres in the timber are parted rather than sawn through**

Above **Modern conversion of oak is done in a mill on a large bandsaw**

Right **Different ways in which a log can be converted**

Structural Grading of Oak

Every human activity in today's modern world seems to be governed by rules and codes, and oak framing is no exception, even though it is based on a technique a thousand years old. Once the initial frame design is completed, it needs to be checked by a structural engineer, who will produce a set of calculations. These will be used to make any adjustments to the frame design, specify the grade of oak to be used and will be included in the building-regulations package. In order to produce these calculations, the engineer needs to know specific information on the strength of the timber. Different types of hardwood (including oak) are grouped together in different classes. In the UK this is done in five strength classes, D30 to D70, which are set out in a document called BS EN 338 (for more information in the UK visit www.trada.co.uk, the website of the Timber Research and Development Association). In the USA timber is classified in three different ways: one for use, one for its manufacturing process and one for its size. For more information visit www.nhla.com, the website of the National Hardwood Lumber Association.

Of course, grading of oak has been done as long as frames have been made by carpenters selecting the best piece of timber for a particular position within a frame. Past experience would have helped in understanding what the predicted long-term performance was likely to be, as the timber dried out *in situ*. Carpenters still need to grade timber visually as they use it because the modern rules do not take into account where the wood is going to be placed in the building. For example, one timber could be good enough to achieve a certain grade but have a knot at exactly the point where a critical joint is placed, whereas a lower-grade timber could be structurally sufficient at the joint and its defect could be placed at a position on the beam where it is supported, and therefore it would not be a problem structurally. The modern rules also make no separation between heartwood and sapwood, although sapwood is prone to insect attack and not durable, so its use should be minimized throughout the frame.

Bottom left **The timber needs to be visually checked for defects by the carpenter before it can be used. This beam was rejected for cross grain**

Below **The grading of oak takes into account many factors; the size of knots and their position on a beam are just a couple. A knot located on the arris of the beam weakens it more than if it is totally enclosed in the middle of the beam**

Shrinkage

Green-oak frames shrink, there is no getting away from it, so predicting shrinkage is of paramount importance throughout the designing and building processes. The end product of what we build should not be seen as the beautifully tight-jointed green-oak frame that has just been erected, but the mature frame after the green timbers have had a chance to dry and shrink. If the oak is selected and made properly in the first instance, this process should be seen as adding character and quality to the frame, not detracting from it. Different ways for detailing the effects of shrinkage will be dealt with later on in this book, so for now I will concentrate on explaining its physical nature.

When a tree is felled it has a very high moisture content (MC), in some cases up to 80 per cent, whereas a dry piece of timber inside a building can have an MC as low as 8 per cent. Initially in this drying process the timber loses what is known as 'free water' from its cell cavities, and once this has gone the timber is said to have reached its fibre saturation point (FSP), which in oak is around 30 per cent MC. At this point the timber starts to lose 'bound water' – that is water contained within the cell walls – and this is when the shrinkage begins. Typically an oak beam in a house could reach an MC of approximately 12 per cent after a number of years but the rate of loss of moisture will affect how much the beam shrinks. As mentioned earlier in this chapter, oak contains longitudinal cells called vessels, which are straw-like in appearance, and if the bound water in the walls of the vessels dries quickly, they tend to shrink a greater amount than if they dry out at a slower, more stable rate.

Right **Green oak begins to shrink when it reaches fibre saturation point, which is around a moisture content of 30 per cent**

Free Water and Bound Water

Free water is that water which is contained within a cell and will evaporate relatively quickly. Bound water is that which is contained within the fibre of the cell wall and therefore takes longer to evaporate.

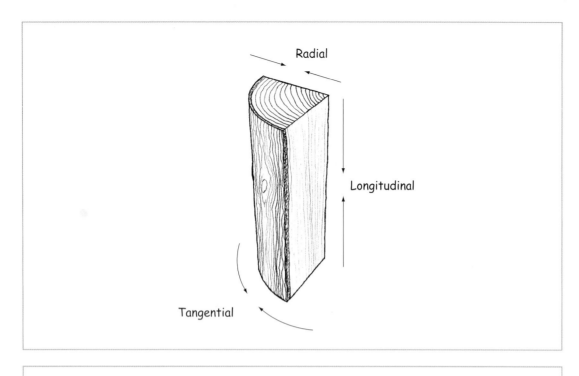

Radial

Longitudinal

Tangential

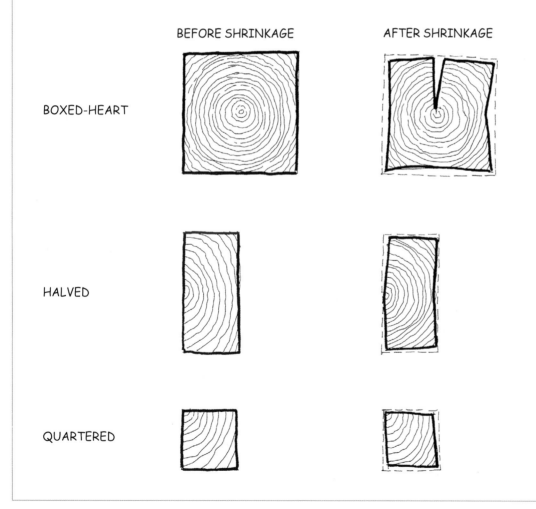

BEFORE SHRINKAGE AFTER SHRINKAGE

BOXED-HEART

HALVED

QUARTERED

So slow drying is good, but that's not the end of the story, because timber shrinks at different rates in different directions, namely radial, tangential and longitudinal. The greatest shrinkage is in the tangential direction (around the circumference, parallel to the grain), and in oak this can be as much as 8 per cent, whereas radial shrinkage (perpendicular to the grain) is about half as much at around 4.5 per cent. Longitudinal shrinkage, on the other hand, is almost negligible at 0.1 per cent, so consequently shrinkage in a frame tends to be localized, as the oak changes in section size rather than length.

The physical effect of these different shrinkage rates on the shape of the timber depends on how it is milled from the log. Boxed-heart timber will start drying faster on the outside, and therefore shrinking quicker, than the wetter inner core, which develops tension in the outer surface of the timber, causing splits (or shakes) to appear. Couple with this the greater shrinkage in the tangential direction than the radial direction, and it causes major splits to radiate out from the centre on each face of the timber. Halved timber, on the other hand, tends to cup around the heart as the outer face shrinks more than the inner one, and quartered timber will become diamond shaped.

Above **The main timbers in this 59ft (18m) oak bridge were connected together longitudinally to prevent shrinkage, which in turn would cause the bridge to deflect**

Top left **Shrinkage is approximately twice as great tangentially (parallel to the grain) as it is radially (perpendicular to the grain). Longitudinal shrinkage is almost negligible**

Left **The effect that shrinkage has on a beam depends on how it was converted. As oak dries a split or fissure appears, radiating from the heart to the closest face. This split releases the tension built up by tangential shrinkage**

Insect Attack

Wood-boring insects tend to be a fairly emotive subject when connected with timber house construction. The present view in the construction industry seems to be that all wood should be treated with some form of preservative in order to protect it against insect attack. But you have to ask yourself: does wholesale spraying of a building with noxious chemicals provide a solution or just cause a greater, more dangerous, long-term problem? Part of the quandary is the general perception that untreated timber will be eaten away in no time, causing the building to have an infestation of insects or even structural collapse. To describe the effects of insect attack on all types of wood is beyond the scope of this book, but what follows will try to set the record straight for oak.

As mentioned earlier, oak tends to be attacked by wood-boring insects in the sapwood only, unless it is kept in a constantly damp environment. The sapwood is softer and contains sugary starches which will attract insects, whereas the heartwood, whilst green, contains large amounts of tannin, which repels them. Once it has dried, it becomes so hard that nothing can eat into it. If the oak is kept constantly damp – for example where one end of a beam is buried in an old, wet wall – the heartwood goes soft again and becomes subject to decay and attack by the likes of the notorious deathwatch beetle. This trouble mainly affects old houses with bad damp problems, but the solution is not to treat the oak with preservatives but to remove the source of the damp; as the oak dries out it will harden and therefore resist further assault.

The main nuisance with a modern, properly detailed oak frame is the sapwood, but this is not always attacked, and even if it is, some people are prepared to live with a small amount of damage rather than treat it with chemicals. Of course, the best way to avoid attack is to minimize the amount of sapwood in the frame, but this is not extremely practical and can be very wasteful in the milling process. Oak is not very permeable and is classified as extremely resistant to preservative treatment, so indiscriminate treatment would be pointless and a waste of time and money. The sapwood, in contrast, can be very effectively treated by using eco-friendly boron-based preservatives. These can be painted on the sapwood locally whilst the oak is still in the framing yard or applied to the completed frame *in situ*. The main types of wood-boring insects to attack oak are as follows:

Ambrosia beetle – These attack the oak either whilst the trees are standing or once they have been felled and left on the forest floor. Unlike most insects they will bore into the heartwood, leaving a blue-black tunnel about the same size as a furniture beetle. They tend to bore across the grain and their tunnels can be distinguished from those of a furniture beetle by inserting a small pin into the hole, which if it sinks in will confirm the presence of ambrosia beetles. They will not attack seasoned timber and will leave the wood once drying begins, so no treatment is necessary.

Furniture beetle – These will attack the sapwood of oak and tend to have a three-year lifecycle, with the adults emerging from May to August. They can be identified by the 'gritty' dust they eject from their short boreholes, which are tunnelled along the grain. Treating locally with a boron-based preservative is very effective.

Powder-post beetle – These will attack the sapwood of oak up to fifteen years after it has been felled, when the starch content is still high. They tunnel along the grain like the furniture beetle but the dust they eject feels floury. Treating locally with a boron-based preservative is very effective.

Fire Resistance

The fire resistance of wood compared to other materials is actually very good, especially for denser timbers like oak. An oak beam will last much longer structurally in a building than, for instance, a steel girder, which, once it deforms under heat, loses all its strength. Oak, like all wood-based materials, is combustible and when it burns it releases gases; some like carbon dioxide are incombustible but others ignite and are burnt off, causing charring of the surface. This charring (or charcoal) is mainly carbon, and its thermal conductivity is approximately a sixth of that of solid timber. This acts as an insulating layer and slows down the rate at which the internal layers of timber burn. Therefore timbers with large cross sections, such as those found in an oak frame, have a greater fire resistance than is usually recognized. The charring rate forms part of the structural engineer's calculation package, when it is submitted for building regulations.

Below **Oak is very resilient to insect attack**

The Structural Qualities of Oak

∧ **Structure of oak –** Oak is a deciduous hardwood with a dense structure. The outer part of the tree is the sapwood which transports the nutrients that are essential to the growth of the tree. The inner heartwood no longer contributes to the growth of the tree; however, it does build up deposits of tannin that make it both harder and more resistant to insect attack. This makes the heartwood more useful for timber framing.

∧ **Cultivation –** Long, straight oaks with no side branches below the crown are the most appropriate for conversion into building timbers, although those trees with a natural curve are useful for constructing crucks.

∧ **Traditional methods of converting logs –** Hewing, sawing and splitting are all traditional ways of converting logs into timber. Each has its benefits and gives the finished timbers a unique appearance.

∧ **Modern conversion –** Most oak these days is converted into beams using a bandsaw of some sort.

∧ **Structural grading of oak –** Oak, along with other woods, is graded depending on its suitability for certain uses. The grading system allows a structural engineer to determine the category of oak that should be used for particular parts of a frame design; however, carpenters should still assess wood visually so that a timber's individual strengths and weaknesses are taken into account.

∧ **Shrinkage –** Three types of shrinkage – radial, tangential and longitudinal – occur, albeit to differing extents, and these alter the appearance of the oak over a period of time as the shrinkage causes splitting (shaking) or distortion as the frame ages.

∧ **Insect attack –** Oak tends to be attacked by wood-boring insects, and this assault tends to take place on the sapwood. Dry heartwood is very resistant to attack.

∧ **Fire resistance –** The fire resistance of wood, particularly oak, is surprisingly high, especially when compared to modern materials.

Right **The structural qualities of oak mean that age adds character to buildings. This modern extension was added to a building that dates from the middle of the last millennium; new additions have been made over the intervening years and the building now has a mixture of styles that marry together to make a cohesive whole**

Chapter 5

Making Frames

Oak-framed buildings are made in an unique way, so special tools and techniques need to be employed. Everything about a frame is on a big scale, from the size of the timber down to the number of peg holes. Large tools are needed to cut the joints and a big workshop for laying out the building. This chapter explains the processes involved in making an oak-framed building.

Tools of the Trade

One of the greatest pleasures in building traditional framed houses is working with green oak. It is a great material to cut, chisel and handle, although it can be a little heavy at times. Carpenters love to collect tools; for some it becomes a great passion and for others just a necessity for the job. But however you view them, they are essential pieces of kit, which, if handled properly, and kept sharp and generally well cared for, become an extension of your hand. Perhaps it's not so difficult to understand why some oak timber framers become almost obsessed with their tools; after all they use them every day and become familiar with each one's character and qualities. Over a period of time, the handle of a tool will acquire its own patina and feels comfortable to the owner's hand, who hopefully has honed the blade to a razor edge.

Because of the size of the timber and the joints cut from it, oak framing requires some particular tools, which are not always readily available from the local tool shop. Oak framing is based on medieval carpentry techniques, so it follows that many of the tools in use during the past few hundred years are still needed today. With this in mind, a good place to start looking for framing equipment is in shops that specialize in old tools, while for modern equivalents, mail-order companies are quite a good source. A list of these suppliers can be found under Useful Contacts, see page 187. The main difference from conventional carpentry tools tends to be the scale. Chisels, saws and mallets, for instance, all need to be much larger than you would find in a conventional carpentry shop. Equipment is also needed to transport the timber and lift it into position, as some pieces can weigh over half a ton. I have listed some of the most essential pieces we use but it is nowhere near an exhaustive list, as that would fill up the rest of this book!

Left **Tools used to construct green oak frames tend to be somewhat larger than those used in normal carpentry**

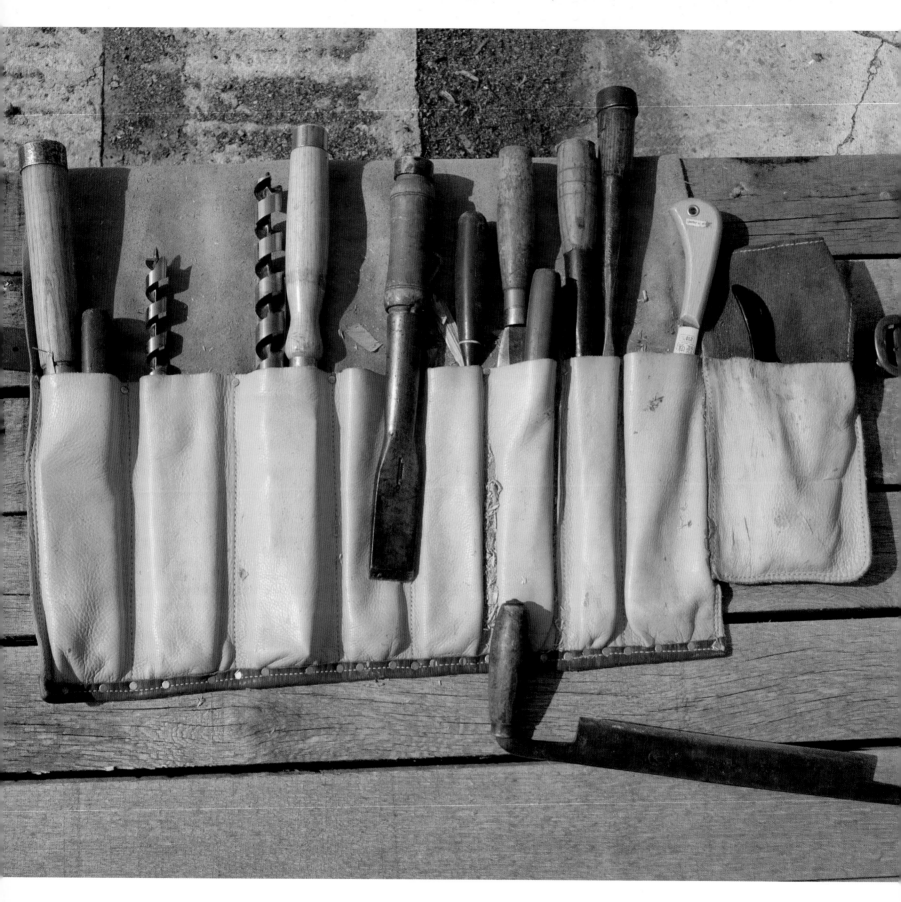

Some important tools:

∧ **Framing chisels** – These are large, tough mortise chisels used for making and cleaning up all manner of joints. The most useful sizes are 1½in (38mm) and 2in (50mm) as they can be used to gauge the width of most common mortises. These can be bought new; the better ones are the socketed type as the handle is less likely to break, but old ones are still favoured because the steel tends to be of a better quality.

∧ **Large mallets** – Mallets sold in tool shops are always far too small to be used with the chisels above, so we tend to make our own. For the head a good piece of heavy, dense hardwood should be used, which will be unlikely to split, such as hornbeam; whereas for the handle something springier should be used to absorb the shock, such as ash. Some people prefer the rawhide mallets, which are encased in iron for extra weight.

∧ **Roofing square** – Also called a framing square, this is invaluable for marking, squaring and measuring angles. The 'blade' of the square – the longer part – is typically 2in (50mm) wide, and the 'tongue' – the shorter part – is 1½in (38mm) wide. Both have a scale in inches or millimetres radiating from the corner. By using Pythagoras's Theorem the square can be used like an ancient calculator, and it is amazing what can be computed by someone who is skilled in using one. The blade and tongue can also be used as a template for marking out mortises.

∧ **Combination square** – A smaller square with an adjustable blade and a spirit level on the stock. These are great for scribing and marking out joints.

∧ **Handsaws** – Two types of handsaw are needed: a ripsaw used for cutting along the grain, and a crosscut saw used for cutting across the grain. Both need to have fairly large teeth, for the fast cutting needed on such large timbers. Modern disposable jack saws combine both the rip and crosscutting action but they are difficult to find with large enough teeth and can not be re-sharpened.

∧ **Plumb bob** – An ancient tool but still the best device for scribing joints. These can be bought but can easily be homemade by melting some lead to form a weight and attaching some fishing line.

∧ **Spirit levels** – Two types of level are needed: a small or 'boat' level for levelling across the face of the timber and a long one for along its length.

∧ **Measuring implements** – It sounds obvious, I know, but several measuring devices are needed: a small steel rule for accurate scribing work, a medium sized 26–33ft (8–10m) tape for most general work, and a long 100ft (30m) tape for setting out and squaring large frames. Measure three times and cut once!

∧ **Chalk line** – Again, a very ancient and simple tool, but nevertheless essential for framing. These are used for snapping lines on the edges of timbers, so that they can be levelled accurately. Japanese ink lines are also very good, but they do tend to mark the timber permanently.

∧ **Drill and auger bits** – Mortises and peg holes in medieval frames were drilled out with hand-held augers, but today they are usually done with a powerful electric drill and separate auger bits. The most common auger bits to use are 1½in (38mm) for drilling out mortises, and 1in (25mm) and ¾in (19mm) for drilling out peg holes. The augers have a threaded screw at the end, called a 'worm', which helps pull the bit through the wood. If you are working out in the woods, with no electricity, a hand-operated brace can be used instead of a drill.

∧ **Chain mortiser** – This is the ultimate mortising device for the professional framer. The chain, which has a series of chainsaw-like teeth, typically comes in the width of the mortise, which for oak is normally 1½in (38mm). Once the depth has been set, it is simply a matter of lining up the chain with a marked-out mortise, and plunging a series of holes.

Arris

The corner edge at which two flat surfaces of a timber join. These are often removed to improve the appearance of the timber and to help prevent it from splintering.

∧ **Hand-held power saws** – Although everything these are able to do can be done by hand, they do speed up the work immensely. The most common sizes are a 9in (225mm) blade for cutting joints and a 16in (400mm) blade for removing waste at the ends of beams.

∧ **Planes** – A number of planes are required for cleaning up joints, and so on, but the two most important for the framer are a rebate plane, which allows you to plane right up to the shoulder of a joint, and a smoothing plane, for flattening a multitude of surfaces.

∧ **Spokeshave** – These are great for taking the arrises off timbers and also useful for cleaning up a whole host of tricky surfaces.

∧ **Drawknife** – These are very useful for removing sapwood from the corners of beams and for making pegs on a shaving horse.

Far left top **Rip and crosscut handsaws**

Far left below **Chalk lines and plumb bobs used for scribing frames**

Above **Power tools used for cutting mortises and tenons**

Below left **Hand planes**

Below right **A spokeshave and a drawknife**

∧ **Large wooden maul** – This is affectionately known as the 'thumper' in our workshop; although in America I believe they call it the 'commander'! Like mallets, these are best homemade and are used for knocking frames together and banging things square.

∧ **The trolley** – Essential for moving timber that would otherwise be far too heavy to lift. It is made with two wheels and a carriage for good manoeuvrability around the workshop, and fits just underneath the height of the trestles so the timber can easily be loaded onto it.

∧ **Trestles** – Not every framer uses trestles, but if like me you want to save your back and work at a comfortable height then they are a must. (In my company it is the apprentices' job to make trestles, and at the last count we had 96!)

∧ **Draw pins or podgers** – These are tapered metal pins, which act as temporary pegs whilst putting frames together in the workshop and erecting them on site. They are made in a T-section, so that they can be easily twisted out of a peg hole when they need to be removed.

∧ **Axes** – I love collecting and using axes, and they are undoubtedly useful for timber framing. Some of the handiest are a side axe for sharpening pegs, a broad axe for hewing and a double-headed axe for the company axe-throwing team!

Far left **Thumpers and podgers**

Left **The trolley**

Above **A collection of hewing axes**

Right **A broad axe (left) and a felling axe (right)**

Ordering Timber

After months of planning, changes in design and production of the final drawings, there comes a time when construction can at last start. The timing of this needs to be planned carefully, so that the ordering of the timber and manufacture of the frame runs in conjunction with the preparation of the site and the laying of the foundations. The timber, for instance, is not – as some people assume – just kept in stock and pulled out when needed, but is ordered in specifically for each job. As you can imagine, no two frames are exactly alike and to keep enough timber in stock to cover all eventualities would be impossible, especially if you consider a large frame could contain up to a thousand separate pieces. It takes time to saw at the mill as well, so at least a month should be allowed from placing a timber order to actually receiving it in the yard.

To be able to order the timber, a cutting list needs to be produced detailing the exact size and quantity needed. This list is created by measuring each member in the working drawings and then recording its description, position, length, section size and cubage (volume). Usually each member is given a unique reference number as well, so that it can be easily tracked once it is being moved around the workshop or yard. A great deal of care needs to be taken in the production of the cutting list, because mistakes made at this stage can prove to be very costly. It is not simply a case of measuring each piece of timber and recording the length, because normally the working drawings do not show the joints, so additional length needs to be allowed for with most pieces. In members such as wallplates and soleplates that appear to be a continual length, a scarf joint needs to be allowed for, and curved timber may need to have its deflection measured. The description of each piece is important from a structural point of view, as certain members such as purlins and joists may need to be cut to a higher grade than soleplates, and if the sawmill has this information they can select the appropriate timber.

The Layout

Once the timber has been delivered to the yard or workshop, work on the frame can finally start. Making large structural frames requires different techniques from those employed in other types of joinery. The timber, for instance, tends to be much larger in both dimension and length, so it is not easily possible to pass it through a planer to make it uniform, as would be common in most joinery. Consequently, when a large structural frame is made the carpenter has to deal with imperfections within each member, as they are likely to be neither very straight nor square. There are four main 'layout' techniques which are commonly employed when building frames and each have many variations. Sometimes different combinations of these techniques are used in a single frame but usually one method is preferred over another.

⋀ **Scribe rule** – This system is commonly used in Britain and involves laying out the various timbers of each frame on top of each other, full scale, in the yard. This technique, called a lay-up or piling, accounts for the irregularities in every timber and is particularly suited to the British style of oak framing. Once each member has been referenced and perfectly placed, the joints can be marked by vertically transcribing the shoulder positions of one timber onto another, with the use of levels and plumb bobs. This system will be described in depth later in the chapter.

⋀ **Square rule** – This system was invented in America around the beginning of the nineteenth century. Its basic premise is that within every out-of-square or bowed timber is a precisely square one. So, for instance, within an irregular 8x8in (200x200mm) post, is a perfect 7½x7½in (188x188mm) post. A ½in (13mm) square housing can then be cut from a reference face to accept a member jointing into the post. The joint of the connecting timber is then cut square and ½in (13mm) longer. In this way it is possible to mathematically work out the length of every timber and the size of every joint without having to lay the timbers on top of each other. It can be used with curved timber but it becomes very complicated.

⋀ **Mapping** – It is possible to account for irregularities in the timber by adding a series of reference lines and level marks to each side and then measuring the deflections at the positions of the joints. These measurements have to be transferred to the corresponding timbers, so both mortise and tenon are cut correctly. It is critical to keep an overall three-dimensional image of the frame in your mind whilst doing this, which can become very complicated on all but the simplest of layouts.

⋀ **Mill rule** – All the timber is planed to a set dimension at the sawmill. For instance, a sawn 8x8in (200x200mm) post can be planed square to 7½x7½in (188x188mm), and so on for the rest of the timber. The joinery can then be worked out simply from a set of drawings, without having to cut housings or lay-up a frame. Unfortunately this method doesn't account for any bows or twists in the timber after it has been planed and it also cannot be used on curved timber which is so apparent in British framing.

Top right **The framing yard has to be well organized to cope with laying out frames full size**

Right **A wall frame in its final assembled position laid out in the workshop**

Below **In square-rule framing, housings are cut at the intersection of each joint, to take out any irregularities in the timber**

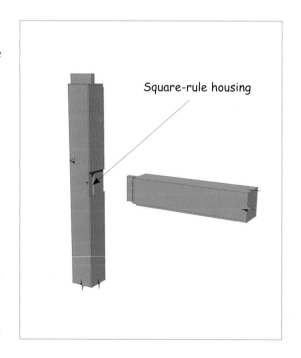

Square-rule housing

Traditionally scribe rule is done in a 'framing yard', which has to be large enough to cope with the spatial requirements of laying out frames, and the storage of timber that is not currently being worked on. Even in the best yards, plenty of time is invested into moving timber around, and this in itself can become quite a big factor in the time taken to build a frame. As the building is built in two-dimensional planes, the sequence of this is very important, even down to the initial stacking and placing of the timber, so that it is easily accessible at all times.

Scribe-Rule Layout

Scribe rule involves laying up each plane of the building in its exact position. This is achieved by stacking the members of a particular frame on top of each other and then transcribing the positions of the joints. The basic principle of scribe rule is that it assumes that each piece of timber is not perfectly square and straight, so by laying out a frame full size, allowances can be made for the irregularity of the timber. To be able to stack the timbers accurately careful reference lines and marks have to be first put on each member. Because most timbers within the frame will be bowing and/or twisted to some extent, these reference lines need to refer to a constant theoretical grid throughout the building. This is where the idea of perfectly flat planes, which bisect the building, comes in very useful. The bends or twists in the timber can move in or out of these planes but the overall geometry will remain constant. The reference lines and level marks placed on the timbers help to describe these imaginary planes, which in turn enable the carpenter to construct the frame accurately. To further assist the carpenter, the planes should coincide with the main setting out points of the complete frame; for instance, under the soleplate, on top of floors or wallplates, on the main cross frames and on the exterior wall frames.

Where these planes cross on the outside of the frame they form origin points or nodes, which in turn can be used to construct a grid on which the building itself can be laid out. The main setting out dimensions of the frame therefore remain constant, and are not subject to any wanderings the timber may be experiencing. This grid is also reflected in the dimensions placed on the working drawings, so for instance a dimension will go from one reference face to another. If the dimensions were put between the timbers, discrepancies would creep in because the timbers are never sawn exactly to the size that they are marked on the drawings. The critical timbers

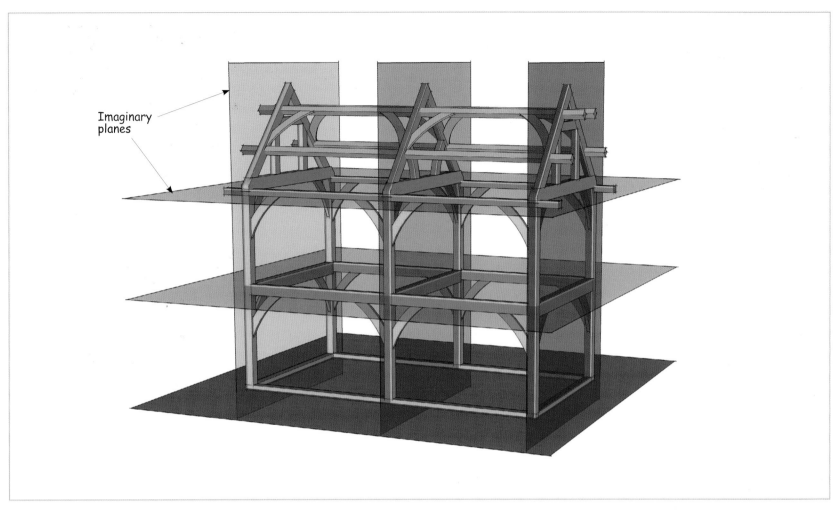

Imaginary
planes

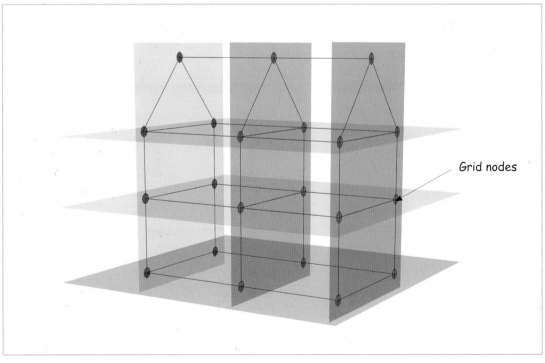

Above **Imaginary planes can be constructed through the building that will create constant reference marks for the carpenter to work from, even if the timber in the frame is bowed or twisted**

Left **An arched-brace truss laid out, with podgers temporarily holding the joints together. The final curve of the braces has been marked and will be cut once the truss is taken apart**

Right **Where different planes cross they create origin points or nodes, from which a grid can be created. The grid can then be used to accurately lay out a frame**

Grid nodes

Above **The primary timbers can clearly be seen in this photo**

Below **The principal rafters in this truss are primary timbers because they are common to the cross frame and the roof frame, whereas the queen posts are secondary timbers because they are only present in the cross frame**

Right **The timbers are faced with a face mark and a face-edge mark before being laid up. The face side is placed uppermost on the trestles and orientated in the right direction. The beams are then packed and wedged to make them level so that they can be scribed correctly**

to be referenced are the ones that are common to more than one reference plane. These are called primary timbers, whereas the ones that are common to only one plane are called secondary timbers. A principal rafter is a good example of a primary timber, as it is common to the roof frame and the cross frame, whereas a queen post is a secondary timber, because it is only used in the cross frame.

Facing

Working to a particular plane means that all the timbers which are part of it have to be referenced and orientated according to that plane. This is done by adding a face mark on the main side, and a face-edge mark on an adjacent side. When the timbers are laid out in the workshop during a frame lay-up, the chosen 'face side', as it is known, is always uppermost on the trestles. In order to face the timbers correctly, they are laid out on trestles and are examined for any defects. These can include structural defects, such as knots, especially where joints may intersect, and any twist or bow. The orientation of the heart is also examined, as its position is likely to affect the shape of the timber. Beams tend to bow up on the heart side of the timber because of the tension released within the wood at the time of sawing. Even within boxed-heart timbers, the heart rarely runs straight and true through the middle, and consequently bows and twists are produced.

As a rule of thumb, the timbers tend to be faced with the position of the heart on the upper and outer side, where the upper is the face side, and the outer the face edge side. In this way all the timbers within a frame tend to be bowing in the same direction. A notable exception to this rule is in the wall frame, where the wallplates are faced with the heart up and out, and the soleplates are faced with the heart down and out. This way the soleplate usually bows down, whilst the wallplate bows up, so that when the wall is loaded with the weight of the roof, the effect will be to flatten them out.

As previously mentioned the face side refers to the imaginary planes through a building. So, in the complete frame, the face side of the external timber will appear on the outside of the building. Therefore the wall frames, roof frames and any gable cross frames will all be made so that the upper face side is on the exterior of the building. Internally, cross frames are faced in connection with the overall design of the building. The timbers on the face side are all set either flush to each other or a set amount apart. Any discrepancy in thickness is all taken up on the non-face side. Consequently the face side tends to look better than the non-face side and is therefore orientated towards the direction where it is going to have most visual impact, for example the main living space. Usually, for the simplicity of framing i.e. running measurements, all the cross frames will face in this direction, except the gables, which will be facing externally. In floor and roof frames, the timber is faced so that it is bowing upwards, so when load is applied they will tend to flatten out. One of the main reasons for facing the timber to the outside of the building is because the heart side of the timber is more durable than the sappier non-heart side. And, laying up and scribing is more predictable if all the timbers are faced and bowing in the same direction.

Referencing Timber

Once the timbers in a section of frame have been faced, they can be orientated roughly into position. To be able to set them out and scribe them accurately, these timbers need to have reference lines and level marks put on them, which relate to the imaginary plane in which they are being framed. In the case of primary timbers these lines and marks need to relate to two imaginary planes, as these timbers will be common to more than one two-dimensional layout. This referencing is critical if the frame is to be made accurately in three dimensions. Take a jowl post in a cross-frame layout, for instance: once it has been faced, two origin points can be marked on. These would be the position that

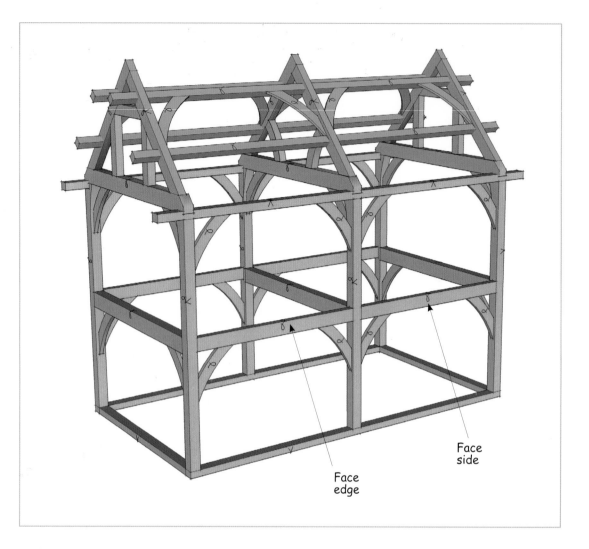

Face edge

Face side

the planes of the top of the wallplate and the bottom of the soleplate would make as they pass through the jowl post. At a point roughly between these marks, a small level is placed on top, perpendicular to the length. If the jowl post is cupped at that point, a little area can be planed flat so the level doesn't rock. A pencil line or a scratch mark is drawn on either side of the level, and a cross is drawn in the middle. This is called the level mark, examples of which have been found on many medieval frames. In this case, the level mark will be parallel to the plane through the cross frame. When the level mark is placed halfway along a beam it has the effect of evening out any twist at the ends. In the case of a twisted jowl post though, it is more useful to put the level mark near the flared end where the head of the jowl connects to the tie beam at its widest point, so it aligns more closely. The twist will be more pronounced at the bottom of the jowl post but it is less noticeable.

When the jowl post is framed in the cross-frame layout, it is wedged so that the level mark is levelled. Then, when the jowl post is framed in the wall frame layout, it is wedged again so that the level mark is perfectly plumb. In this way, it is known that the jowl post has rotated through a perfect 90°, regardless of how square or twisted it is, and the overall geometry of the frame is kept true.

Apart from the level mark, two chalk lines need to be snapped, one down the face side, and one down the face edge. These chalk lines visually represent the imaginary planes through the cross frame and the wall plane, only set back a parallel amount from the origin point. These enable it to be laid out accurately, even though it may be quite bowed. Taking the jowl post as an example again, a 1in (25mm) mark is drawn on both sides of the corner between the face side and the face edge side, at the origin points (the top of the wallplate mark and the bottom of the soleplate mark).Two chalk lines are then snapped between these points. They could be placed 2in (50mm) down, or even on the centreline of the jowl, if that makes it more convenient to mark the joints. These reference lines are usually only chalked on the primary timbers. On some it is useful to chalk more than two, for instance a tie beam needs to be referenced to its lower edge, even though it is faced towards its top edge. This is because tie beams are faced to indicate the direction of bend, not the working plane (bends should always point upwards in horizontal beams). A useful line to put on is one 1½in (38mm) up from the tie beam's bottom edge, taken from the outside of the wallplates, as this is normally the depth that the tie beam is dovetailed over the wallplate. This can be adjusted to suit dovetails made at a different depth.

Laying Up Frames

The next stage of the operation is to lay out one complete plane, such as the wall frame, so that it can be scribed. The order in which the lay-ups are constructed is always much debated. Some framers believe that it is better to set the soleplates out first, as they set out the overall dimensions of the building, whilst others believe it is better to do the cross frames first or even the wall frames. The decision usually comes down to the practicalities of moving timber around, and which pieces are most accessible! To produce a lay-up the timbers are accurately stacked on top of each other, so that the joints can be marked on by transferring the irregular profiles of the lower one onto the upper one, and visa versa. The timbers can be stacked several layers high, so many pieces can be scribed in one go. Practically this can become difficult because you have to be very careful not to accidentally knock any of it out of position. Scribing as well becomes less accurate the higher you go. Normally a frame is laid up twice: generally once to scribe and cut the primary timbers and then to scribe and cut the secondary ones.

There are two common methods for positioning the timbers in the correct place. The first one, called 'lofting', involves drawing a layout full size on the floor, if it is clean enough! If it is not possible because the surface is rough or uneven, boards can be fixed to the ground so origin points can be marked on them. The first set of timbers, usually the largest, are then placed over the lofted drawing and

Far left **Every timber within the frame is faced in a particular direction. The face side of the external timbers is always on the outside of the building. Internally the timbers tend to be faced to point the same way**

Below left **This jowl post is referenced with a 'level mark' which is parallel to the plane through the cross frame. Two origin points are marked on, which in this case are the positions of the top of the wallplate and the bottom of the soleplate. Two reference chalk lines are then snapped, one on the face side and one on the face edge side**

Below **A cruck frame being 'lofted' in the workshop**

levelled along the reference lines, across the level mark, and also in relation to each other. This is done by packing and wedging the individual timbers up, on blocks off the floor. Their position is then checked very carefully, by transferring origin points marked on the lofted drawing to the corresponding timber by using a plumb line. Once the position of the first set of primary timbers has been established, the position of the second set, those that will joint into the first set, is marked on. The second set can then be laid above the first set, and then packed so that they can be scribed accurately. Lofting an accurate drawing on the floor takes time to do, and laying out the timbers above it is not straightforward and is also

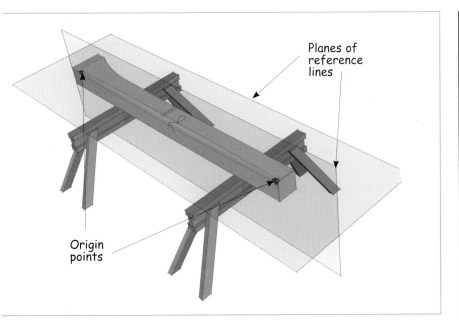

Planes of reference lines

Origin points

time consuming. Lofting comes into its own when repetitive frames need to be made, especially if they are complicated or use very curved timber. Although it takes time to plot the original drawing, this can be used as a template for each frame, eliminating any measurement errors. It is particularly useful when making cruck frames, where each blade tends to have a slightly different curve, yet needs to be set out so the overall geometry of the roof, for instance, stays constant.

The second, more common method of frame layout, is to do it by setting out the timbers by measurement. Taking the cross-frame layout as an example again, firstly, jowl posts are placed roughly in position, with the face uppermost and levelled. Reference marks need to be put on each one, which will not be lost by the cutting of any joints. Typically a 12in (300mm) reference mark is placed below the top of the wallplate mark and above the bottom of the soleplate mark, on the chalked reference line. These are done because the original marks will be lost when the joints are cut in the ends of the jowl posts. The jowl posts are then moved until they are parallel to each other and the correct distance apart, using the reference lines, so there are no discrepancies from any deflections. Next they need to be squared, which is done by checking the diagonal measurement between the origin points,

and making any adjustments where necessary. Working from the plans, the position of the next set of timbers, which will be laid on the jowl posts, can be marked on. As with lofting, the second set of timbers are carefully placed upon the first set (the jowl posts in this case), on the correct joint positions. The members that lie above the origin points, in this case the soleplate and tie beam, need to be level both along the reference line and across the level mark. Both ends of these timbers will theoretically fall through the same distance when they are scribed so their upper face surfaces end up flush with the upper face surfaces of the jowl posts.

For the floor beam that lays across the middle of the jowl posts a different approach needs to be taken. This is because the bows in the middle of the frame can be much more extreme than those at the ends near the origin points. If the floor beam is scribed when level but finishes in a frame out of level, because of the bow, the shoulders of the joint will be wrong. By packing each end to the same height, a vertical drop scribe can be done, whereby each end falls by the same distance to its final position, regardless of its longitudinal level. The floor beam still needs to be levelled across its level mark, so it can be framed correctly in the floor lay-up and, as before, this is done by wedging. Once the floor beam is finally fitted in position, the

cross frame reference lines on the jowl posts need to be transferred to it, so the correct plane can be established when laying up the floor.

Laying out frames by either method takes a lot of practice to do well and mistakes are easy to make if care and attention are not taken. The carpenter needs to have a clear picture in his or her mind of how all the components in the frame are interacting, and the ability to think in three dimensions is essential.

Scribing

The theory of scribing one piece of timber onto another so a perfect joint can be cut is quite simple, it is only when it is applied over a complete frame that it becomes complicated! There are many different ways to scribe joints and every carpenter seems to have his or her favourite method. In the end it's about mating irregular pieces of timber together to form perfect joints. What follows is one method I use.

As described earlier, the timbers in the first layout are stacked on top of each other in their correct positions. Both the lower set and the upper set have been levelled or packed correctly, with the faces of the timber pointing upwards. Even though the middle of the timbers may bow upwards, the reference chalk lines describe a flat plane through each set of timbers. Now if the upper set could drop magically through the air to the level of the lower

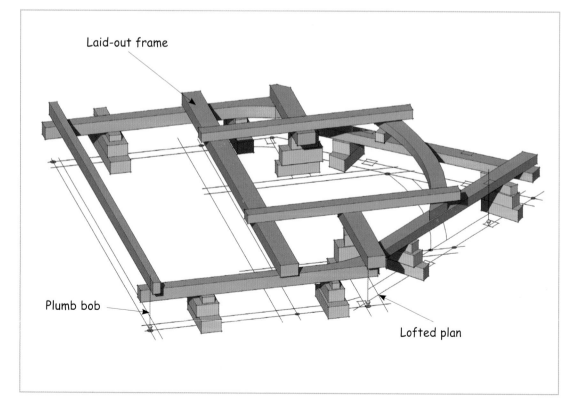

Laid-out frame

Plumb bob

Lofted plan

Left **'Lofting' a frame involves drawing it out full size on the floor. The timbers are then placed above the drawing, levelled and then scribed. This technique is very useful when making repetitive frames or crucks**

Above right **Primary timbers being 'laid-up' and squared into the correct position**

Right **Timbers are set out by measurement and then packed and levelled for scribing**

Far right **Timbers are scribed at the joints by vertically transferring the shoulder positions of one timber onto another using levels and plumb bobs**

set, and, where the timbers met they fused together, a perfect joint would be formed. This is what scribing tries to achieve, by transposing edges of one timber onto another, where the joint is going to be formed, by transferring them vertically.

Take an out-of-square floor beam, in a cross frame, that is jointed into an equally out-of-square jowl post, as an example. Both are primary timbers and would have been positioned during the first layout, with the floor beam stacked on top of the jowl post and packed to the right distance off it. A plumb line is slid along the top edge of the floor beam until it is just touching one arris of the floor beam and one arris of the jowl post. If both timbers were perfectly square the plumb line would connect fully with both faces of the timber. The top face of the floor beam, in its final position, will be flush with the top face of the jowl post, so the amount that the upper arris on the jowl post deviates from the plumb line is marked on the upper arris of the floor beam, in other words, directly above. How this deviation is measured depends which way out of square both timbers are. This process is repeated for the other three edges on the floor beam, and the four points joined up will define the plane of the jowl post. At the same time, the plumb line is used to mark the position of the floor beam on the edge of the jowl post, where the mortise will be cut.

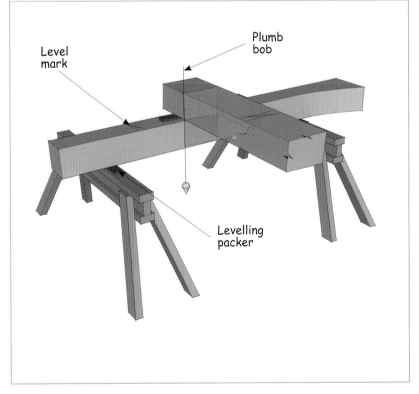

Marking Joints

After all the primary timbers have been scribed in a particular layout, the job of marking out the joints can fully begin. This is a highly skilled process, in which the carpenter has to make many decisions. Some of these will depend on the location of the joint and how it might interact with other joints in the area, and others will depend on how the joint is working in the frame. Using the scribed lines, part of the joint is marked while the timber is still in its layout position, to reduce the risk of making a mistake (which tends to be very costly when dealing with large pieces of oak)! After the preliminary marking has taken place, the stacked timber is dismantled onto separate trestles. The timber can now be turned over, so the marking on the underside can be completed. Both the mortise side and the tenon side are marked at the same time, to ensure that they mate perfectly. Due to the level of skill involved, the more senior carpenters tend to do most of the scribing and marking, but when it comes to cutting the joints, everyone pitches in, so they have to be labelled very clearly.

Cutting Joints

When every joint is marked, the cutting can begin. Each individual frame will normally be made by a pair of carpenters, so one will start chopping out the mortises while the other will work on the tenons. These will be roughed out with hand-held power tools, and finished off by hand. At this stage, peg holes are marked and drilled through every mortise. The size of these holes varies depending on what type of joint they are holding together, but typically they are either ¾in (19mm) or 1in (25mm) in diameter. Once all the chopping, planing and drilling is completed, the frame can be reassembled.

Below left **A joint being scribed with a plumb bob**

Below **A tenon is marked on the joint by using a combination square**

Left **Once the tenons have been marked, they are 'roughed out' with a hand-held circular saw**

Below left **The joints are cleaned up using hand tools**

Below right **The peg holes are drilled through the mortises before the frame is reassembled**

The Secondary Layout

Unlike the first layout when the primary timbers were stacked, this time they can be fitted together in their newly formed joints. This is where the 'thumper' comes in handy. The same process of levelling and squaring is performed again and if any of the joints don't fit perfectly, they are adjusted. Sometimes it is necessary to pull the frame tight, with the aid of ratchet straps, so that everything fits and it is perfectly square. The centres of the peg

holes, previously drilled through the mortises, are marked on the tenons by pressing in the worm of the auger bit, so at a later stage the joint can be draw-pegged. The attention given to this process is vital, because the next stage of the operation is to scribe the secondary timbers. In a similar method to the first layout, the positions of the secondary timbers are marked on the assembled frame. The secondary timbers are examined for defects and faced like the primary ones were before. This time though, they

Left **Wind braces and purlins 'laid-up' in the workshop ready for scribing**

Right **Once the joints have been cut after the first lay-up, the timbers are reassembled square and level and temporarily podged together. The secondary timbers are now laid on in order to be scribed**

Below **The tenon peg hole is drilled slightly offset from the mortise peg hole before the frame is reassembled**

are only given a level mark, as the chalked reference lines won't be needed. The reason for this is because they will be scribed in the same way as the floor beam as described in the first layout.

The plumb lines now come out again and a second round of scribing begins. The joints are marked out as before and the frame is dismantled so they can be cut. Whilst the frame is apart, peg holes are drilled through the mortises of the secondary joints and also the tenons of the primary joints. The centre of the mortise has already been marked on the tenon, but now the peg hole is drilled offset from this point, nearer the shoulder of the joint. When the joint is reassembled, the peg hole in the mortise won't quite line up with the one in the tenon, so when a tapered peg or podger is knocked in, the holes will pull together, thereby tightening the joint in the process.

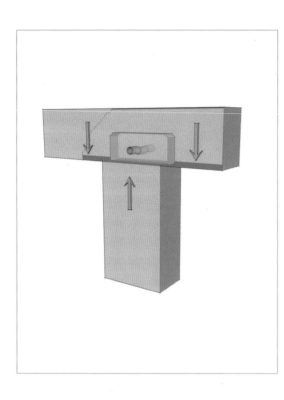

Draw-pegging

Draw-pegging or draw-boring is a technique that is used to tighten joints. In the workshop it helps to keep the frame fixed and square when scribing secondary timbers, and when the frame is raised it pulls joints tight permanently. The mortise timber is pre-drilled with peg holes prior to final assembly and when the tenon is fitted, the centre of the mortise peg hole is marked on it. The tenon peg hole is then drilled offset towards the shoulder of the joint. When the joint is assembled the two holes will not quite line up, so when a tapered peg is driven in, the tenon timber is pulled tightly into the mortise timber. This pulling action continues after the joint has been pegged, and so helps to minimize the effects of shrinkage as the timber dries out. The amount of draw, or offset, given to the tenon peg hole depends on the size and type of peg that will be driven into the joint. As the peg is driven in, it will bend as it follows the route of the peg holes. So if too much draw is used the peg, or even the tenon, could break. The smaller ¾in (19mm) pegs are more flexible than the 1in (25mm) pegs, so the draw tends to be greater, typically between ⅛in (3mm) and ³⁄₁₆in (5mm), whereas for a 1in (25mm) peg the draw should only be between ½in (1mm) and ⅛in (3mm). The dowel or turned-type pegs, with a sharpened end, should have less draw than the handmade drawn pegs, which are more flexible.

Final Assembly

With all the joints cut in the secondary timbers, the frame can be assembled for the last time before the final erection on sight. The same method of squaring and levelling takes place again, but this time the primary joints can be podged tight, which aids the process. Once all the secondary joints have been checked, and the frame is perfectly square, they too have their tenons marked for later draw-boring. This is a very satisfactory moment, because for the first time the complete frame is laid out in the workshop. It is not always possible to

Above **Peg holes on the tenons are drilled slightly nearer the shoulder of the joints than the peg holes on the mortises. When the joint is pegged up the two holes try to align, thereby forcing the mortise and tenon tight together**

Right **In the final assembly the tenons of the secondary timbers have their peg holes marked when the frame has been checked for being square. Each member of the frame now has its carpenters' mark chiselled into it**

scribe a frame in two layouts, so sometimes a third layout is needed. This occurs particularly when a frame contains many interconnecting secondary timbers. Before the frame is finally dismantled, each member in the frame needs to be numbered.

Carpenters' Marks

As each timber in the frame has been scribed individually, its position is unique and not interchangeable. Therefore it has to have an exclusive number or 'carpenters' mark' to distinguish it. The numbers are historically based on the Roman numerals, probably because of the ease of chiselling them in. There is no set way of doing these marks, so individual groups of framers tend to develop their own system. What's important, however, is that all the carpenters working on a frame together use the same system! What follows is a simplified explanation of how we do it.

Left **Carpenters' marks need to be chiselled on every timber so that they can be correctly located and orientated in the final assembly**

Below **The position, style and number of the carpenters' mark tells the framer exactly where the timber belongs in the building when the frame is erected**

As we tend to start by making the cross frames, the upper or face side of those members are marked first. Each cross frame is given a sequential number, so, for instance, in a four-bay building they will be numbered one to five. The wall frame on one side of the building will be called the 'straight' side, and on the other, the 'curly' side (the reasons for this will soon become clear). When the first cross frame, or cross frame **I** is laid out, the members on the straight side of the building have a single **I** chiselled in them, with a normal mortise chisel, whereas, on

the curly side of the building, they have a single **C** chiselled in them, with a large gouge. Members such as floor beams, which span from one side of the building to the other, will have both – that is an **I** on the straight side and a **C** on the curly side. Subsequent cross frames will be marked in the same way: for example, the third cross frame will have **III** marked on the straight side and **CCC** on the curly side. The process is further complicated by placing the marks in a particular position or a different size on each member, to identify its correct orientation

within the frame. Studs, for example, which could be placed upside down, always have their marks at the top end, whereas braces have their marks at the bottom end. The practice is continued in the wall, floor and roof frames. So for example, in a roof frame, the purlins and windbraces that connect into the principal rafter **CC**, will also be marked **CC** on the face side. When the frame is erected on site, a skilled carpenter should be able to pick up any member and know its exact position and orientation, just from the carpenter's mark.

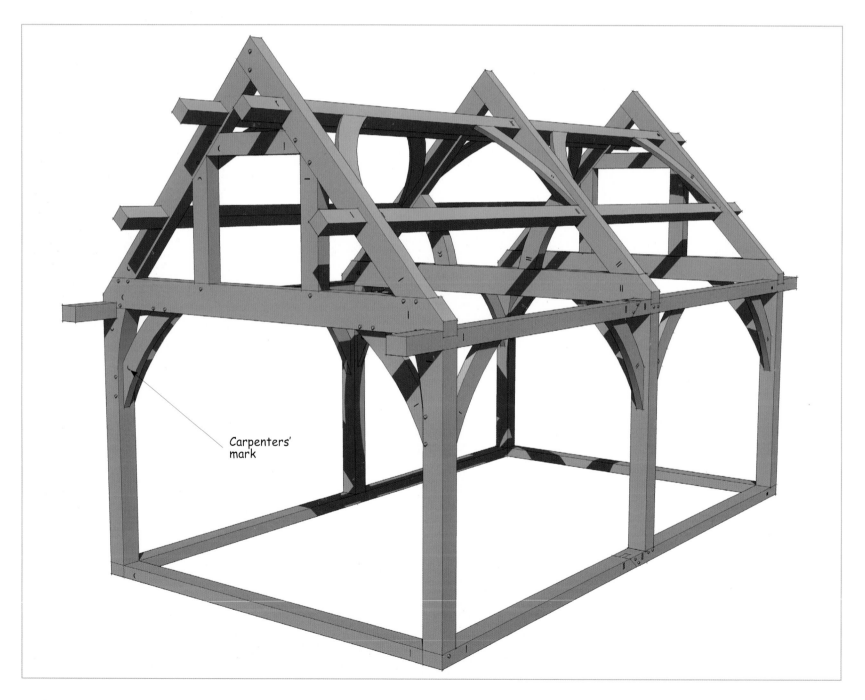

Carpenters' mark

A Note About Wall Frames

Wall frames need to be laid out in a slightly different sequence to the other frames. The reason for this is that the wallplates and soleplates tend to be very long continuous members (usually the length of the building), made up of several individual sections. Before the wall frame can be laid out for scribing, the plates have to be scarfed together to form a continuous length. First, each section of timber that will make up the complete run has to be faced. The wallplates are faced so that the heart is pointing up and out of the building – in other words the timbers will be bowing upwards. The soleplates are faced with the heart pointing down and out of the building – in other words the timbers will be bowing downwards. (They are faced like this in anticipation that the weight of the building, once fully erected, will flatten them out.) After they are faced, reference chalk lines are snapped so the scarf joints can be marked out. The position of these scarf joints requires careful planning because the lower half of each scarf joint has to support the connecting upper half. This is particularly important with the wallplates. The lower half of the scarf should be positioned near a primary post, which will support the joint.

Far left **A frame showing the position of carpenters' marks**

Below **Wallplates and soleplates are scarfed together before they are laid-up for scribing**

Bottom left **The lower part of the scarf joint needs to be placed near a primary post for sufficient support**

The direction in which the joint is cut is also important for the erection sequence, as this dictates whether the end of each section has the lower or upper part of the scarf. When the frame is erected (see Chapter 6: Raising Frames) the first section of plate needs to have the lower half of a scarf joint at the end where it connects into the second section. The second section then has the upper part of the scarf joint where it connects into the first section, and the lower part of the scarf where it connects into the third section, and so on. In this way the plates can be craned up and dropped onto each other in succession as the frame is erected. Once all the plates have been correctly scarfed they are podged together to form a continuous length, after which the wall frame can be laid out for scribing.

Above **A joint being cut in the workshop**

Left **A floor beam with a bevel-shouldered tenon for additional support**

Joints

The jointing decisions made by the carpenter during fabrication are critical to the structural integrity of the frame. They are the weakest point of the frame, so a lot of care and attention needs to be taken, not only when deciding what joint to use, but also in the making of it. Joints need to be skilfully cut if they are to perform well, but this is a task the experienced carpenter relishes.

The first consideration in making any joint has to be the quality of the timber where the joint is going to be formed. Joints should not be made where there are any structural defects in the timber, such as knots over a certain size, as this will further weaken an already stressful point of the frame. When timber is laid out for framing it is positioned to avoid any defects at the joints. If this is not possible, because for instance the member is too short to be slid one way or another, then it must be rejected and replaced with something more suitable.

Joints should have well-scribed shoulders and be clearly marked out, before any cutting has commenced. It is important to maintain consistency in marking throughout the frame, because the carpenter who cuts the joint is often different from the one who has marked it out.

Most joints need to fit snugly, but not so tight that they are difficult to assemble. This is usually achieved by making the tenon slightly narrower than the mortise. The reason for this becomes clear when the frame is erected on site, as a certain amount of flexibility in the frame is needed to fit all the members.

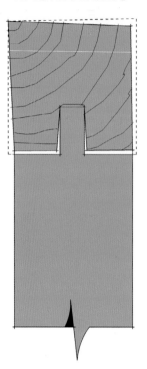

BEFORE SHRINKAGE

AFTER SHRINKAGE

MORTISE CUT
TOO SHALLOW

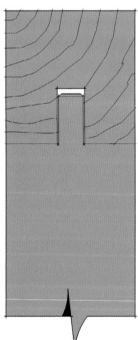

BEFORE SHRINKAGE

AFTER SHRINKAGE

MORTISE CUT
DEEP ENOUGH

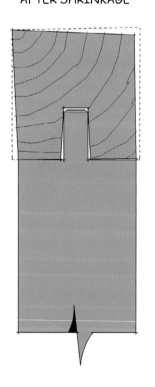

If the joints are made too tight it is sometimes impossible to spread the frames apart. Also, some buildings may take weeks to make in the workshop, and by the time they are erected some of the timber may have moved slightly. A very tight-fitting joint made in the workshop becomes an impossible joint to fit on site. On the other hand joints do not want to be too imprecise because they will stop working effectively, so striking the correct balance is essential.

Mortises need to be made deeper than tenons, to stop the tenon 'grounding out' on the bottom of the mortise as the frame shrinks. The length of the tenon, as it is cut along the grain, will not be affected by shrinkage, whereas the mortise, which is cut across the grain, will be. As the timber shrinks, the depth of the mortise will reduce, so if the tenon is not cut short enough to allow for this it can push the joint apart. Typically the mortise needs to be made ½in (13mm) deeper than the length of the tenon.

As a carpenter, one constantly has to weigh up what is going to be the best joint to use, for any given situation. It's not just the type of joint, but its size and position. If on the one hand you make a tenon larger to improve its strength, on the other, the mortise will also have to be made larger; potentially weakening the timber it is housed in. There is a constant balancing act involved in joint design between improving the strength of one piece, whilst not weakening the other. Joints may also be concentrated at certain points around the frame, creating 'clashes' which have to be avoided. Quite often these are not easy to see, because although they appear at the same area, they will often be in different planes, only to be discovered when a different section is laid out. Most of these joint clashes can be avoided with foresight, but a three-dimensional mindset helps. What follows is a selection of the most commonly used joints in our workshop, but there are many variations of these that are not listed.

Left **Mortises need to be cut deeper than tenons, otherwise when the frame shrinks the joint can be pushed apart by the tenon 'grounding out' on the mortise**

Facing page:
Top left **Basic mortise and tenon**

Top middle **Barefaced tenon**

Top right **Stopped tenon**

Right **Splined bevelled-shoulder tenon**

Bottom right **Bevelled-shoulder tenon**

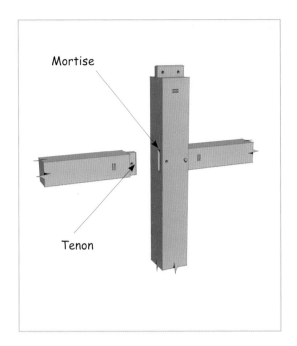

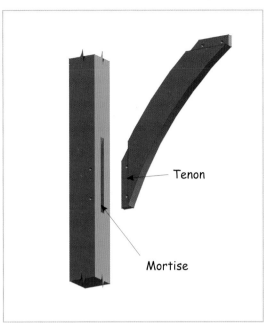

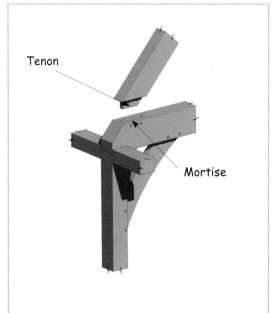

∧ Basic mortise and tenon – This is the most common joint, and is used extensively throughout most frames. The usual thickness of the tenon is 1½in (38mm), although this is sometimes increased to 2in (50mm) on large timbers or those acting more structurally. The tenon is cut out of the end of a beam, leaving either one or two shoulders that will be in contact with the mortise beam.

∧ Barefaced tenon – A barefaced tenon is cut on one side only and so produces only one shoulder. These are commonly used in braces where the timber is not usually thick enough to produce two shoulders and a reasonable-sized tenon. One end of the brace tenon is cut perpendicular to the shoulder, so the mortise is not undercut.

∧ Stopped tenon – If the tenon is placed near the cut end of the mortise timber, then a stopped tenon is used, so the end of the mortise doesn't break out due to horizontal shear. These are often used on the principal-rafter-to-tie-beam tenon, where the back of the mortise on the tie beam has to resist the outward thrust from the tenon on the principal rafter.

∧ Bevelled-shoulder tenon – If a load-bearing beam is jointed into the side of a post, then a bevelled- or toed-shoulder tenon should be used. An angled shoulder line is cut, projecting past the scribed shoulder line in the beam and a corresponding housing is made in the post. The full width of the beam can now transmit the load into the housing, instead of just relying on the strength of the tenon. The bevel should be at least 1in (25mm) deep, although this depends on the size of the members, to allow for any shrinkage that may occur in the post.

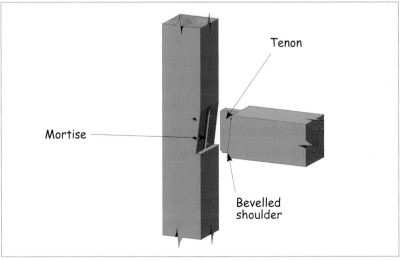

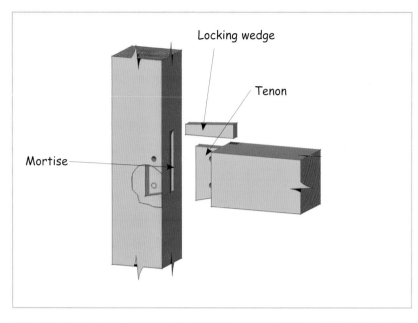

Locking wedge

Tenon

Mortise

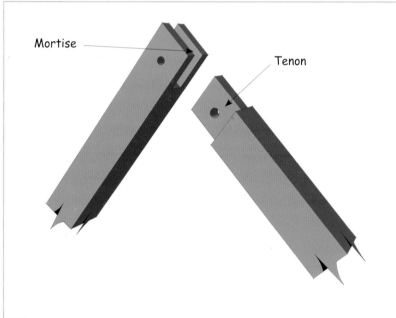

Mortise

Tenon

Dovetailed tenon – If a joint is in tension – in other words the stress is trying to pull it apart – a dovetailed tenon can be engaged to lock it together. Extra length is given to the mortise, so the joint can be assembled, and then a wedge is driven in above the dovetail to lock it. Unlike a compression joint, where the force is transferred into the shoulders of the tenon, a tension joint needs longer tenons if it is to resist horizontal shear.

Bridle joint – A bridle joint is an open mortise and tenon joint, used to connect rafters and principal rafters. The tenon is the full depth of the mortise timber, and works in compression, as the weight of the roof pushes the two rafters together.

Tusk tenon – A tusk tenon is a load-carrying joint, for beam-to-beam connections. It is a housed-in mortise and tenon, so the load is carried by a substantial depth of timber, rather than relying just on the tenon. The upper shoulder of the tenon is angled back to reduce the size of the mortise. The best position for the tenon is along the neutral axis of the mortise beam, between the areas of tension and compression.

Bridle scarf – Scarf joints are used to join two or more pieces of timber together in their length, to form one long piece. The bridle scarf is commonly used to join wallplates because it works very well at opposing both horizontal and vertical forces.

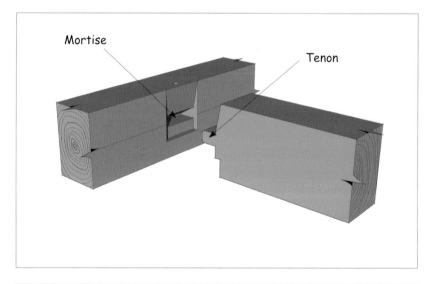

Mortise

Tenon

Top **Dovetailed tenon with locking wedge**

Above **Bridle joint**

Right **Tusk tenon**

Bottom right **Bridle scarf joint**

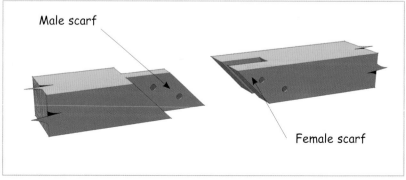

Male scarf

Female scarf

∧ **Stop-splayed and tabled scarf** – This is a very strong scarf joint that is held by driving wedges between the two cut, inclined planes of the timber, forcing them together, although pegs are also added for extra strength. Scarves should generally be supported near the lower half of the joint, because, however good they are, they are still not as strong as a whole piece of timber.

∧ **Dovetailed lap** – This simple lap joint is usually used between the tie beam and the wallplate, to hold the walls together. If the wallplate tries to move away from the tie beam, the dovetail joint will keep tightening, resisting any movement. The depth of the dovetail is usually marked on the tie beam at the cross-frame layout stage, as it will set out the position of the wallplate.

∧ **Spline joints** – A spline is a separate piece of timber used to join two other pieces together. It acts like a large tenon and is placed in mortises in the receiving timber. These are very useful for crowded joint situations, for instance where the floor and wall beams connect into a primary post. A spline running in one direction connecting the wall beams can pass another, at a different height, connecting the floor beams. Less wood is removed from the post because large tenons are not needed, and pegging is simplified because none of the holes are covered up. Another position for using splines is on jointed purlins. Again less wood is removed from the principal rafter, as normally a tusk tenon would be used, but it is also a much safer joint to assemble because the trusses don't need to be spread as far, as the purlins have no tenons on the end. Splines are usually made out of air-dried oak stock that is 1½in (38mm) thick.

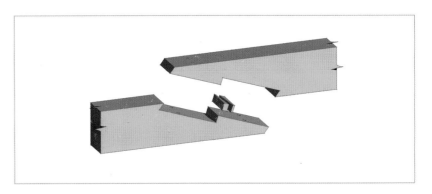

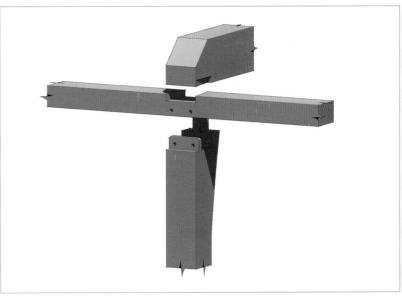

Top **Stop-splayed and tabled-scarf joint**

Above **Dovetailed lap joint**

Left **Spline joints**

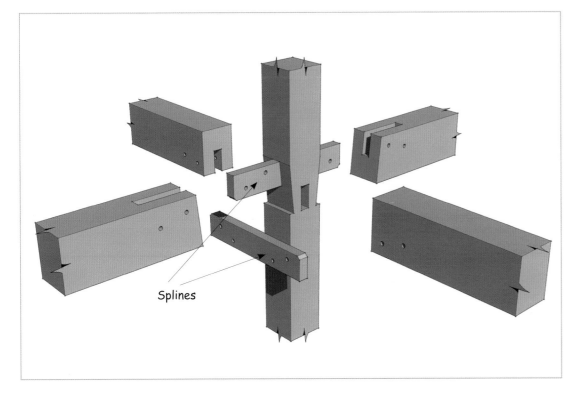

Splines

Cutting Curved Timber

Curved timber is what makes British-style frames so unique. This means that the cutting of it to its final shape is very important. It is normally delivered from a timber supplier either 'in the round' or sawn 'through and through' (slabbed), for final processing by the carpenter. The timber is first sawn to the correct thickness before the shape of the curve is marked on, normally by using a large flexible stick. The shape must follow the grain of the timber, otherwise when it dries it has the potential to develop a structural fault. Smaller pieces of curved timber, such as windbraces, and wallbraces, are usually cut out using a big bandsaw, but larger pieces, such as cruck blades, are cut using a chainsaw.

Making Pegs

Pegs (also known as 'tree-nails') hold each piece of timber and, in turn, each frame, together. They are obviously very important to framers, who end up driving in thousands every year. They are put under huge stress as they are forcefully knocked in with a large mallet or lump hammer, and are required to hold the joint structurally sound for hundreds of years. Consequently, they have to be made properly, from well-selected timber, and because of the numbers involved, this can be a relentless task, which in our workshop falls to the apprentice.

The stock used to make pegs needs to be from a fast-grown oak for strength, and very straight-grained because any short grain that is present in the peg

Above **Curves are cut so their shape follows the grain of the timber as in this example of a hammerbeam frame**

will cause it to snap. It should also be green so it will split easily. The first stage in producing pegs is to cut a suitable log or beam to the finished length of the peg. A grid is then marked on the end grain of one face of the timber, the size of which will depend on the diameter of peg being made. Using the grid, the timber is then split into square blocks, called 'blanks', slightly larger than the finished diameter of the peg. The timber is split using an implement called a froe, which basically consists of a long blade with a handle attached perpendicularly to it at one end. The froe is positioned over one of the grid lines and knocked very hard, with a big wooden maul (or thumper), causing the oak to split. Split or cleft timber is much stronger than sawn timber because the split will follow the grain, separating the fibres instead of cutting through them.

Above **Froe and drawn pegs**

Left **A log being split with a froe to make peg blanks**

Froe

A metal-bladed tool that is used to split wood. The depth of its blade is given a wedge shape that forces the timber apart as the froe is struck. The direction of the split can be determined by using the handle as a lever.

The blanks are given their final form by one of two methods. A traditional style called a drawn peg is fashioned by hand-shaping the blanks on a shaving horse. This is done using a drawknife to gently taper the blank into a roughly octagonal shape. A template, with holes of varying sizes drilled in it, is used to check the diameter of the peg as it is being produced. These pegs are very good for draw-pegging and don't need any further sharpening.

The second method, resulting in a regular dowel peg, is to drive the blanks through a peg maker. This simple device is basically a removable sharpened metal cylinder mounted on a tripod. The internal diameter of the cylinder will be either ¾in (19mm) or 1in (25mm) to match the pegs' finished diameter. The blanks used to make dowels should be stacked to air-dry for a while, which will ensure that the finished peg doesn't shrink excessively once in the frame. Once the blank has dried sufficiently it's placed over the cylinder of the peg maker and knocked through with a wooden mallet until the perfect peg drops out of the bottom. The dowels need to have a sharpened end, if they are to be use for draw-pegging, and this is created with a side axe.

Above **Peg blanks being knocked through the 'peg maker' to make dowels**

Left **Peg being made on a drawhorse with a drawknife**

Making Frames

∧ **Tools** – A wide range of tools is required to build a timber frame, including: framing chisels, mallets, various squares, saws, plumb bobs, levels and measuring devices – not to mention the numerous power tools for drilling and cutting timbers, and a whole host of planes for tidying up the joints. Draw-pins or podgers are used to assemble parts of the structure temporarily.

∧ **Ordering timber** – A cutting list details the exact size and quantity of timber needed, and timbers are ordered specifically to meet the structural requirements of each part of the frame.

∧ **The layout** – The system of layout – scribe rule, square rule, mapping or mill rule – depends on the type of frame that is being built. This book deals in large part with the scribe-rule system that is the mainstay of British-style oak framing.

∧ **Facing timber** – The orientation of the heart and the cut of the timber, along with any structural flaws such as knots, determines the side of the timber that will face out.

∧ **Referencing timber** – Once the primary timbers in a section of frame have been faced, they can be placed in position. A level mark will keep the timber correctly positioned, while two chalk lines are used as theoretically straight reference lines.

∧ **Primary layout** – The next stage of the operation is to lay out the primary timbers, either by 'lofting' or by measurement. This enables the timbers to be scribed and the joints to be marked.

∧ **Plumb scribing** – Once timbers are laid out the edges of one timber are transcribed onto another by transferring them vertically.

∧ **Cutting joints** – To start with, joints will be cut with hand-held power tools, before being finished by hand. Peg holes are marked and drilled through each mortise.

∧ **The secondary layout** – Timbers are laid out again but this time they are fitted with their joints, and any adjustments are made. The peg holes are marked on the tenons so the joint can later be draw-pegged.

∧ **Draw-pegging** – This helps to tighten the joints and minimize the effects of shrinkage. The pegs are offset to encourage the timbers to be drawn together, although the extent of the offset depends on the demands of the joint.

∧ **Final assembly** – Once the joints are cut in the secondary timbers, the frame can be assembled for the final time. It is squared and levelled again with the primary joints podged tight. The secondary joints are then checked, and their tenons are marked for later draw-boring.

∧ **Carpenters' marks** – Each timber's position is unique so it must have a relevant carpenters' mark to distinguish it from the others. 'Straight' and 'curly' marks are used to denote the side of the building that the timber belongs to, while the number of marks indicates the relevant cross frame. Further specifications such as orientation are identified by the size and position of the mark.

∧ **Joint type** – A huge variety of joints are used in order to meet the demands and withstand the stresses of certain parts of the building. The main ones used are varieties of mortise and tenon: barefaced tenon, stopped tenon, bevelled-shoulder tenon, dovetailed tenon, bridle joint, tusk tenon, bridle scarf, stop-splayed and tabled scarf, dovetailed lap and spline joints.

∧ **Pegs** – Pegs are relatively small but still an incredibly important part of the frame. They are made from straight-grained fast-grown oak and split along the grain in order to be able to withstand the immense stresses placed upon them.

Chapter 6

Raising Frames

After the hard work of designing and making the frame comes the excitement of raising it. This chapter starts by looking into various types of foundation, and how to organize a safe building site. It then goes on to give examples of different methods of raising frames, and the sequence in which they are erected.

Foundations

In tandem with manufacturing the frame in the workshop, the groundworks need to commence on-site, and be completed by the time the frame is ready to be erected. The foundations of the building are often paid little attention, I suppose because they are largely unseen, buried beneath the ground. However they are in fact the most important part of the building; the initial building block on which everything else stands. Many medieval buildings collapsed in the past, not because the oak frame failed but because there were inadequate foundations. An oak-framed building can be designed and built to last many hundreds of years, so it is imperative that the foundations are designed accordingly.

The correct type of foundation to use will depend on a number of physical and environmental factors. The function of the foundation is to transfer the structural loads from the building into the subsoil below, without any unacceptable movement during the lifespan of the building. The average domestic oak-framed building, clad in normal materials, is classified as a lightweight structure, so the main consideration in designing the foundation is the soil condition, and other local factors such as the presence of trees. Soils with a large amount of gravel in them are subject to only a small amount of movement, whereas soils that retain a large amount of water, such as clay, can change volume near ground level. During different weather conditions the ground can expand or contract, and this can seriously affect the foundations, especially if they are not dug deeply enough.

Subsoil

The subsoil is the layer of soil that lies beneath the topsoil but above the bedrock. The type of subsoil is crucial not only in terms of the solidity of the foundations but also other important factors such as water run-off.

Left **Soleplates being set out in their correct positions on the foundations**

Trees that are growing close to a building can also affect the foundations either by causing shrinkage in the soil, during periods of dry weather, or through physical contact with the roots. Trees close to a building or over a certain height can significantly increase the necessary depth of a foundation. Even if a close tree is felled before construction starts, the foundations may still need special design. Once the roots stop draining the ground, the soil can expand or 'heave' in wet weather. The design of foundations should be left to specialists, normally a structural engineer. This is especially the case if the plot is on a sloping site and retaining walls need to be built, or if a cellar is part of the construction.

The two most common types of foundation both involve digging trenches around the perimeter of the building. With traditional strip foundations they are partially filled with concrete to form a firm footing and are built out of the ground with brick or blockwork, whereas trenchfill foundations, as the name suggests, are filled completely with concrete. Most self-builders these days use trenchfill because although the material cost is higher, it's much faster and so saves on the cost of labour. Also, strip

foundations have to be set out more accurately, so there are fewer margins for error. Providing there are no ground problems the trenches are normally dug 3ft (1m) deep and either 18in (450mm) or 2ft (600mm) wide, the width of the digger bucket. The wider trench costs more in materials but gives more tolerance when setting out the walls, especially if they are constructed with a wide build-up, such as a faced cavity wall.

A foundation is needed anywhere where a load needs to be transmitted to the subsoil. This will normally be around the external walls and gables but also at points where internal load-bearing posts reach ground level. These are catered for by either digging further trenches across the building or using isolated pad foundations – 2ft (600mm) square holes filled with concrete. With strip foundations, the soleplates and load-bearing posts need to be fully in contact with footings below – which sounds obvious, but all too often they are set out to the external skin of the frame. If a block-and-beam floor is going to be used then internal piers need to be built under any load-bearing points down to the foundations as most floors of this type are not capable of carrying much load.

Below left **Trenchfill foundations are costly in materials but are fast to construct**

Below **Strip foundations have to be set out more accurately than trenchfill foundations**

Right **Piled foundations are used when the ground is poor**

Below right **A soleplate set out over a piled foundation with a block-and-beam floor slab. Piles were used because of the proximity of this mill house to the river**

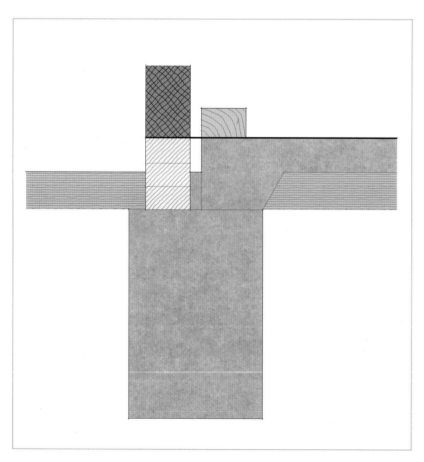

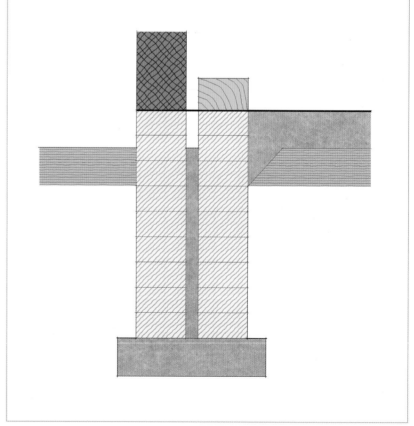

Pile and Raft Foundations

On problem sites with poor ground conditions, the alternatives to strip or trenchfill are either a raft or piled foundation. Normally a concrete floor slab is poured after the foundations have been completed, but with rafts the concrete for the foundations and slab are poured at the same time, with the whole structure reinforced with a mesh of steelwork. The raft acts like a snowshoe spreading the load over a wide area, absorbing any movement in the substrata. This type of foundation is particularly suited to lighter buildings resting on poor soils.

Where deep foundations are required, the best and cheapest alternative is to use piled foundations. Instead of a trench, a series of holes are drilled until firm ground is reached, and then filled with concrete. The piles are then connected with a ground beam of steel and more concrete, which can then be built on in the normal way. Piles are very useful when trees are close to the building or soil is particularly unstable, for instance near a river.

The Site

Building sites never look particularly pretty but it's important that they are well organized and safe. Safety should be the number one concern in any self-builder's mind. Having a cluttered site, with rubbish strewn around, will make everyone's life more dangerous. Apart from the safety aspects, an untidy site is an inefficient site. Every task will take longer to do and this will affect morale. Before the frame arrives, the site should be cleared and an area of hard standing laid, particularly if the weather is poor and the ground soft. This can be done with a load of scalpings purchased from the local quarry. This not only helps to keep the site clean and stable underfoot, but will also stop any heavy plant from sinking up to their axles in mud.

An oak-framed building of a reasonable size contains at least 20 tonnes of oak, and many contain much more. This takes up a lot of space and will probably be delivered on an articulated truck. These are large vehicles, and they need good access on and off the site. Of course, it's not always possible to get near the site, in which case other options need to be explored. For example, the timber can be off-loaded in a friendly farmer's yard, and transported to site with his or her tractor and trailer or delivered in several loads with a smaller lorry. I have learnt over the years that it is always possible to deliver the timber, but sometimes it takes a huge amount of planning.

Scalpings

The offcuts of stone that result from the stone being dressed for building work. Almost anything can be described as scalpings so the quality of the material you receive can be extremely variable.

Hard Standing

A hard standing is a hard surface on a building site on which vehicles may be parked without fear that they will sink into the ground, or otherwise become unstable.

Right **An oak frame delivered to site can take up a lot of space; in this case the timber had to be unloaded directly onto the structural slab**

Below **A hard standing is needed on-site so that the crane and other heavy vehicles do not sink**

When the frame is unloaded, the timber needs to be spread out as much as possible. Clear access to the site and room for a crane should be left on the hard standing. In some circumstances a crane can be brought on site first, and used to unload the lorry, simplifying the operation. The timber needs to be spaced so that no dangerous stacks are produced, and identifying each piece is that much easier when it comes to raising the frame. Finally, when the timber and crane are in place the area should be made secure, so unauthorized people are kept away.

Cranes

Using a crane is by far the safest and quickest way to raise a frame. They are big, intimidating pieces of kit to the uninitiated, but to trained personnel an absolute godsend. They come in all sizes from the 6 tonne babies to the massive 100 tonne (and bigger) giants. The weight refers to how much they can lift but this diminishes the further away the load is from the crane. So for instance, a 25 tonne crane may only be able to lift a load of half a tonne, at a distance of 65ft (20m) from it. The size of the crane, and the space required for it, will therefore be calculated by how far it needs to reach, with a certain load, around the building. Using cranes is no laughing matter and only fully qualified personnel can go near them. If asked, crane operating companies will usually pay a site visit to advise on the correct size of crane and to check the access. It is possible to engage a crane company to do all the lifting, signalling and slinging (strapping the timber to the hook of the crane) to erect a frame. Most professional framing companies have qualified personnel to do the signalling and slinging and hire in the crane and operator to do the lifting.

Top **Once the oak has been unloaded it needs to be spread out round the site**

Bottom **Part of a truss being lifted in with a crane**

Far right top **Cruck being lifted whole with a gin pole**

Far right bottom **Shearlegs like these are a variation of a gin pole**

Gin Poles and Shearlegs

Frames can of course be raised by hand, without the use of a modern crane. It's not only a quieter method but also it's very satisfying to overcome the laws of gravity with just poles, rope, pulleys and a lot of sweat. If a frame is not too large, most of the timbers can literally be carried to the frame and just lifted into position. Once the frame exceeds a safe working height or when the timbers become unmanageably heavy then it is best to use a traditional lifting device.

A block and tackle can provide the mechanical advantage to enable heavy weights to be lifted. The problem is that the top block needs to be placed higher than the final position of the member you're trying to lift. One way of achieving this is to use a 'gin pole', which basically is a straight vertical wooden pole, restrained by guy ropes. A block and tackle is securely fixed to the top of the pole. At the same position restraining guys are also fixed. At least two are needed pointing backwards, opposite to the way the pole will lean, and one pointing forward. It's then raised vertically into position above the object that needs to be lifted. Ideally once erected the top of the pole needs to be at least 3ft (1m) taller than the highest lift. The base of the pole has to be firmly fixed to stop it moving. As the timber is lifted the weight of it will stretch the back restraining guys and make the pole lean forward. To compensate for this they need to be attached to a separate block and tackle system, so the lean in the pole can be adjusted for. The front restraining guy stops the pole from falling backwards and helps stabilize the whole apparatus. Obviously the anchor points for the guys need to be very solid and if they are not exactly in the right place extra guys will be needed. The system works very well, but takes quite a long time to set up just for one lift. Before even attempting to use a gin pole the load needs to be calculated, so that the correct gear can be used.

A variation of the gin pole is 'shearlegs'. They consist of three poles lashed together to form an upright letter A. The rigging is much the same as the gin pole, except only one back guy is needed. This can be an advantage when suitable anchor points are hard to find. The guys need to be perpendicular to the plane of the shearlegs otherwise they become unstable. Traditionally shearlegs were used to raise the masts on ships. It goes without saying that using both these devices can be very dangerous in the wrong hands, and they should only be used by skilled and trained operatives.

Above **A talk on site safety**

Below **Scaffolding should be placed around the foundations before the erection of the oak commences**

Far right **Safety nets like these should be spread across the frame once it is up to wallplate level**

Site Safety

On-site construction is inherently dangerous, unless proper precautions are taken from the outset. Even self-builders are legally responsible for ensuring that on-site risks are minimized. A risk assessment should be carried out at the planning stage, identifying any potential risks and how they can be avoided. Only trained and competent people should be employed, and they should be provided with the correct equipment and protective clothing. Welfare is also important, so washing and toilet facilities need to be supplied. Most fatalities on building sites are the result of falls, so adequate protection should be factored in from the beginning. In order to put a frame up safely full scaffolding should be erected around the perimeter. Scaffold towers should also be used internally within the frame, and once the frame has reached the height of the wallplates, safety nets should be spread between them in each bay. It is beyond the scope of this book to detail every health-and-safety measure needed to build a house, but further information can be found from the Health and Safety Executive at www.hse.gov.uk or the Occupational Safety and Health Administration at the US Department of Labor at www.osha.gov

Raising the Frame

I find raising the frame the most exciting and pleasurable part of the whole construction. It can be the culmination of months of hard graft and is the reason why I first fell in love with my job, and why I still enjoy it so much today. It is always fascinating to see the joints sliding perfectly together, and the frame eventually taking shape in its three-dimensional form. The speed at which the frame shoots up, sometimes taking just one day, is very exciting, but there's always a sense of relief that everything fits properly. It can also be a very emotional experience for the home owners we are building for, who perhaps see the realization of their dream appearing magically before their eyes. But a word of warning: erecting oak frames is a very skilled occupation and should only be attempted by professional timber-frame carpenters who are trained and competent.

Setting Out the Soleplates

The starting point for erecting any frame is setting out the soleplates. If these are perfectly level and square, there should be no need to check to rest of the frame as it is erected. It amazes some people that this is the case, but if the frame has been made properly, as described in the last chapter, it will pull itself plumb and square as the braces are fitted. First the slab needs to be checked for levels, in the positions where the soleplate will lie. Discrepancies should be recorded on the foundation plan, so the plates can be packed level later. Depending on the foundation design, the next stage of the operation is to roll out a damp-proof course (dpc) or radon barrier under the soleplate positions. The soleplates can now be laid out, and packed up to the correct height, usually with pieces of ply or slate, under the main load-bearing points, and the joints pegged up. Finally, the diagonals are measured to check everything is square, and tweaked where necessary (the 'thumper' comes in useful for this). After the frame has been completely erected, any gaps under the soleplates should be backfilled with lime mortar or a weak mixture of sand and cement.

Damp-Proof Course

An impermeable layer that is incorporated at the lowest construction level to prevent the detrimental effects of rising damp.

Assembly or Rearing

There are two distinct methods for raising frames, which depend on whether the frame is of the post-and-truss or cruck type. Usually a frame is assembled piece by piece like a large jigsaw puzzle, but with crucks, it is typically necessary to rear the whole cross frame in one section. All the members of the cruck frame are laid out horizontally with the feet of the blades above their respective mortices in the soleplate. The frame is completely pegged and reared vertically into position, where it needs to be temporarily shored until it can be attached to the wall-frame members. Traditionally this would have been achieved with a block and tackle, large poles and many men, but nowadays it is more often than not done with a crane.

Top left **The brace tenons in this wall frame have to be lined up with the mortises in the wallplate as it is craned into place**

Bottom left **A truss is gently lifted onto a cross frame**

Top **The soleplates have to be level and square before the erection of the main frame can begin**

Right **Temporary shoring is used to secure the posts before they are tied in by the frame**

Assembling up to the Wallplate

The most common place to start assembling a frame is at one of the gable corner posts. This is craned into position and fixed with temporary shoring to keep it stable. Following this the rest of the gable cross frame is erected up to wallplate level, the joints being secured with podgers and the whole thing being kept stable with more shoring. The jowl post of the next cross frame is now erected as before, and then connected back to the previous cross frame with the edge beam running at upper-floor level. The rest of the second cross frame is now erected up to wallplate level, and finally the opposite edge beam is fixed to complete the connection between the two frames. This is an important stage of the raising because one bay is now joined completely at first-floor level. This provides a stable platform for the rest of the frame to be connected to. It is essential that no timbers are left out in the rush to erect the frame, because it is impossible to fit most of them retrospectively. It is especially easy to miss out secondary timbers such as braces or studs, or to put them in the wrong position. The carpenters' marks, so carefully put on in the workshop, become invaluable during the erection, as they not only identify each member but also show its correct orientation.

Right **The erection sequence has to be carefully planned so every timber is safely put up in the right order**

Below **Jowl posts waiting to receive the wallplate**

At each stage some of the joints need to be finally pegged before their holes are covered by connecting timbers. Many joints, though, are left temporarily podged, until the frame is completely finished. This is because a certain amount of flexibility is required in the frame to locate the joints. At times certain podgers have to be removed, so that the frame can be spread slightly to accommodate a linking member.

After the first bay is completed, the subsequent cross frames and connecting wall frames are built until every member connected to the first floor and below is erected. If there are joists in the frame, they are fixed at this stage to provide a safe deck for working on the upper levels. The wallplates and any secondary timbers joining into them are then craned into position. The wallplates, being long, continuous pieces of timber are usually made up of several pieces. The sequence in which these are erected depends on the direction of the scarf joints (see Chapter 5: Making Frames). The first section of wallplate should have the lower half of the scarf joint, whilst the second should have the upper half, so the second can be laid on top of the first. This alternation of scarves continues along the length of the wallplate.

Raising Trusses and Purlins

After the wallplates have been erected, work can start on the roof. The way the trusses are erected ultimately depends on their design and the way they are connected into the cross frames. But generally, for reasons of safety, each truss is pegged together on the ground and raised in one lift. The joints in the cross frame, in most cases, are designed in such a way that a truss should be able to be lowered vertically onto the receiving upright tenons. Mating up a whole series of mortise-and-tenon joints in one lift can sometimes be tricky, but once again the thumper usually comes to the rescue. After the first pair of trusses has been erected, the purlins and windbraces in that bay are fixed to stabilize them. This can be tricky and dangerous if the purlins are of the jointed type, as the trusses have to be pulled apart to locate them. The use of spline joints in this situation is a much safer option. Once the first bay is completed the rest of the roof is raised bay by bay.

Left top Framing pins or 'podgers' are used to temporarily fix the frame together

Far left bottom A section of roof is sometimes craned in whole

Left bottom Trusses like these are normally craned in whole

Above The truss is pegged together on the ground before it is lifted in

Above right The joints in the cross frame have to be made so that the truss can drop vertically onto them

Right Spline joints are often used on purlins because it is easier and safer to erect them. The trusses do not need to be spread so far to locate the purlins, as is the case with tenoned joints

Pegging Up

Finally, the moment arrives when the last timber in the frame has been raised, and everyone can stand back and admire the completed structure. This is not the end of the job though, as there will probably still be hundreds of pegs to drive in and the fixing of the common rafters to be done. Driving a peg into a draw-bored joint is a skill in itself. If there is not enough taper on the peg or too much draw on the tenon, the peg or tenon could simply break. A novice carpenter working on a frame soon learns the best way to 'peg up', though practice makes perfect. After driving in a couple of hundred pegs you feel like you've had plenty of practice! They are usually driven in from the face side, until they project through the joint. Both ends of the pegs are then sawn off to approximately ½in (13mm) from the face of the timber. They are not cut flush, so there is a chance to drive the peg in further, if needed, at some point in the future. I also think they look better if they are a bit proud. There is a lot of satisfaction to be had pegging up a joint. What was loose while the frame was being erected becomes tight and firm. All the hard work put into scribing the frame in the workshop now bears fruit as the joints are squeezed together and take their final form. As the pegging continues around the frame, the structure begins to stiffen up, so by the time the task is completed, it should be rigid. It should also be square and plumb, as the braces will triangulate the frame to the correct position.

Far left **Pegs are normally driven in from the face side right through the joint**

Below far left and below left **Pegging up can be hard work**

Below **Different sized pegs are used depending on the joint and position**

Right and below right **The joints pull very tight after they have been pegged because they are draw-bored**

Below **Dowel pegs need to be sharpened with a side axe before they are driven in**

Fixing Rafters

The common rafters are manhandled into position and securely fixed into the purlins and wallplate. The typical way of doing this is by using long galvanized nails (ordinary nails will be eaten by the tannin in the oak) or square pegs, or a combination of both. The rafters are set out parallel to each other and are normally spaced between 12in (300mm) and 18in (4500mm) apart. Depending on the eaves detail, they will either stop at the wallplate or sail past to over-hang the roof. Hips and valleys are fixed in conjunction with the rafters, as setting them out is easier with something to work off. Jack or cripple-jack rafters (those which fix into hips and valleys) are offered up prior to being fixed so they can be saw-carved to fit perfectly. It's hard work heaving rafters into position but at the same time good fun. As the sweat increases on the brows of the carpenters, so does the good-natured shouting across the roof, accompanied by the incessant banging of their hammers. Perhaps this is because the job is drawing to a close and there is an 'end of term' feeling in the air.

Topping Out the Frame

As the framing crew pack up their tools, there is one job still left to do, and that is to 'top out' the frame. In my company we send the newest member of the team to the highest point on the roof, to nail up an oak branch. This ceremony is accompanied by much shouting and cheering until the new recruit has completed the task. This ancient custom originates from Scandinavia, where it was used to celebrate the completion of a building, and to appease any gods or spirits that were dislodged from the trees used in the structure. It is also supposed to bring good luck to the frame and to the carpenters who worked on it. Whatever it signifies, the proceedings are usually followed by copious amounts of beer!

Top left **Rafters are manhandled into position**

Above **Jack rafters are 'saw-carved' so that they fit perfectly**

Below **An oak branch is used to 'top out' the frame**

Saw Carving

Saw carving is a technique by which a hand saw is run between two imperfect surfaces, until they mate exactly.

Raising Frames

∧ **Foundations** – Normally either strip foundations or trenchfill foundations will be used to transfer the load of the building to the subsoil. Alternatively, pile and raft foundations are used on sites with poor ground conditions.

∧ **The site** – This should be well organized and safe, with a hard standing area and access for trucks or cranes if possible. Timber should be well spaced to avoid a potentially dangerous stack and to aid identification.

∧ **Raising the frame** – Either by hand or with a crane. This starts by setting out the soleplates. A post-and-truss frame is assembled piece by piece while a cruck frame will usually have a cross frame raised in one section.

∧ **Starting the assembly** – This is normally started at a gable corner post, followed by the rest of the gable cross frame up to wallplate level. Joints are secured with podgers before the jowl post of the next cross frame erected. A connection is then made with the edge beam running at upper-floor level and the second cross frame is now erected to wallplate level, before the opposite beam is fixed.

Further cross frames and connecting wall frames are built until every member connected to the first floor and below is erected.

∧ **Raising trusses and purlins** – Trusses are usually pegged together on the ground, and then lowered onto the tenons. Purlins and windbraces are fixed to stabilize the trusses and the roof is raised a bay at a time.

∧ **Pegging up** – Once the entire structure has been raised there are hundreds of pegs to drive in before the common rafters are fixed. This tightens the joints and secures the structure of the frame.

∧ **Fixing rafters** – The common rafters are manhandled into position and securely fixed into the purlins and wallplate. The typical way of doing this is by using long galvanized nails (ordinary nails will be eaten by the tannin in the oak) or square pegs or a combination of both.

∧ **Topping out** – An oak branch is nailed to the highest part of the building to appease gods and spirits dislodged from their trees and to bring good luck to the frame and the carpenters.

Below **The site was very muddy due to the extreme weather conditions**

Bottom left **The SIPs are connected by an insulated spline fitted into a groove that runs around the panels**

Bottom right **The panels are fixed to the frame on a counter batten and strip of vapour barrier to form a service void**

Oak Frame and Structurally Insulated Panels

The Castle Hill frame was constructed out of green oak on a very muddy site in Warwickshire, UK during the winter of 2013/2014. The clients wanted to have a high performance building with low energy usage, so it was decided to use Structurally Insulated Panels (SIPs) for the building's envelope. SIPs consist of polyurethane insulation sandwiched between two layers of OSB board. The panels are connected together by means of an insulated spline (or joining piece) inserted into a groove on the side of the panels. This creates a continuous layer of insulation with almost no cold bridging. The panels are put not only on the walls but on the roof as well. The overall effect of this blanket of insulation is to create an extremely thermally efficient building.

The panels are made off-site in factory conditions, so they have to be designed to fit the frame correctly. The Castle Hill SIPs were designed at the same time as the frame in a 3D CAD package. If both are designed concurrently they should fit together like a glove. The beauty of off-site construction (much like the frame) is the time saved on-site, which can be a massive cost benefit.

After the frame was erected, during some atrocious weather conditions, work started on the SIP panels. First a counter batten was fixed to the outside of the frame. This is so a service void can be created (see detail on page 154), and strips of vapour barrier were then fixed to that. This makes it possible to create a continuous vapour barrier

after the SIPs are erected. The SIPs sit on their own soleplate, which was first levelled to the frame. After that it was a question of fitting the ground floor panels one by one, fixing them together and to the frame. The first floor panels were fitted in the same way and after only a couple of weeks all the walls were complete.

Normally roof panels are too large to be manhandled, so a crane is employed to lift them up. To make the raising more efficient, whole sections of them were joined together on the ground. This was done over several days until the back half of the roof was made, and it could be craned up in one go. The process was then repeated for the front of the building. Once all the SIPs had been securely fitted to the frame, using very long screws, it was completely covered with a breather membrane. This not only helps the passage of water vapour from the inside of the building to the outside, it also acts as a waterproof layer to protect the SIPs while the roof tiles and weather boarding are being fixed. The whole operation took four weeks from start to finish to create a watertight shell for this 4,850ft^2 (450m^2) building.

Above left **The ground and first floor panels were individually fitted to the frame**

Above **The roof panels were made up in sections on the ground**

Below left **Whole sections of the roof were fixed in position with the aid of a crane**

Below **Once all the panels had been securely fitted to the frame, the whole structure was covered in a breather membrane, which would help to protect it. The picture shows the membrane in the process of being put on**

Below **The situation in which an oak frame is to be built places unique demands on the building envelope**

Below right **An external envelope that is being fitted onto an oak frame**

Far right **SIPs on the roof of an oak frame**

Chapter 7

The Building Envelope

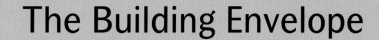

Raising the frame is certainly a huge milestone in the building process, but before the house-warming party can begin there is a long road left to travel. The erection of a structural oak frame will provide the skeleton of a house on which the external envelope needs to be hung. The frame will support the roof, walls and floor but it needs an outer skin to provide it with protection from the elements. The choice of which material or system to use can be complicated, and ultimately will depend on a number of different design factors. There is no clear answer pointing towards one particular method as being 'the best way', as factors such as orientation, noise and aesthetics greatly influence the design.

The Outer Skin

I have said 'outer' skin because in most cases it is preferable for the envelope to be fixed to the outside of the frame and not in-filled between the frame as a medieval house would be. The reason for this is to optimize heat and energy conservation, because it is difficult to achieve really good U values (see box on page 148) and airtightness by using an infill panel. A green-oak frame will shrink away from an infill. If, on the other hand, all of the materials that make up the external envelope are applied to the outside of the frame, shrinkage should not be a problem. In terms of thermal insulation, materials that are applied in a complete layer rather than in between timbers are also much more effective. Timbers exposed both internally and externally can act as a 'cold bridge' for the passage of heat from the interior to the exterior. They can also allow movement of air through cracks which develop as the frame dries. With an external envelope it is much easier to control the airtightness of the building, and so limit the amount of heat lost.

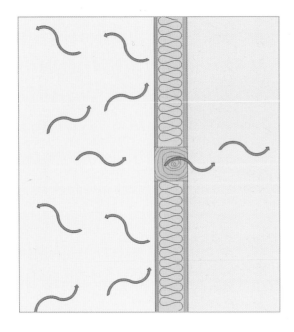

Above **Timber exposed both internally and externally in a frame can act as a 'cold bridge' for the passage of heat**

Right **Traditional infill panels like these at Mary Arden's Farm, Wilmcote, Warwickshire, UK need to be maintained as the oak shrinks**

Traditionally, as the frame shrank around a wattle-and-daub infill panel, it would have been 'lime washed'. This watery paint would run down any newly formed cracks and seal the frame. It would have been reapplied every year, especially when the frame was new. Today a few oak-frame manufacturers do supply modern infill panels as part of their buildings. Most systems involve the use of expanding gaskets to cope with any potential shrinkage. Whether infill panels will survive future building regulations is yet to be seen.

U Values

The thermal insulation of materials used in the building's envelope, such as the walls and roof, is calculated in terms of the rate of heat flow through it. Known as the 'U' value, it is the measurement of the heat energy transmission rate in watts, through one square metre of the construction per degree of temperature difference: W/m^2K. The smaller the figure, the better the insulation. The wall of a building, however, can be made up of several different materials, each with its own U value. It is not possible to calculate the total U value for the wall if the individual figures are added together, as mathematically you end up with the incorrect answer. So, first each component's U value has to be converted to its thermal resistance or 'R' value. This is simply done by dividing 1 by the U value, and therefore the higher the R value, the better the resistance to the transfer of heat. The R values of each material can then be added together to give a total resistance to the passage of heat through the wall. The inverse of this figure (dividing 1 by the resulting value) then gives the total U value for the wall. All materials can achieve the same U value; it just depends on how thickly they are applied.

Sustainability

Designing a sustainable home should be a number one priority for any housebuilder. Using an oak frame is very environmentally friendly but that's not the end of the story. The amount of energy a home uses over its lifetime will far exceed the amount of embodied energy used in its construction. In order to move towards becoming a low-carbon economy we need to reduce our energy consumption and increase the use of renewable resources. So-called 'zero-heating' or 'passive' houses are already being constructed, and although most current houses built in Britain are well below this standard, the trend is slowly moving in that direction. Good thermal design is not just a question of applying ever-increasing amounts of insulation but looking at how the building as a whole is performing. The basic principles for designing low-energy buildings can be summarized as follows.

∧ **Orientation** – The building should be positioned to gain as much access from the sun as possible, and shelter from prevailing winds.

∧ **Minimizing heat loss** – Heat generally is lost in equal amounts through the fabric of the building, the glazing and ventilation. It is possible to improve all three areas by increasing the level of insulation, specifying the correct area of glazing and type of glass and making buildings more airtight.

∧ **Passive solar design** – Glazing needs to be orientated correctly to gain benefit from passive solar heating during the wintertime. Thermal mass within the building should be located correctly to act as a heat store (see Chapter 8: Finishing).

∧ **Energy-efficient appliances** – Low-energy heating and lighting systems should be used such as photo-voltaic cells to produce electricity, solar panels to produce hot water and low energy bulbs for lighting.

∧ **Ventilation** – After the building envelope has been made more airtight and is better insulated, ventilation needs to be carefully controlled if heat loss is to be minimized (see Chapter 8: Finishing).

∧ **Choice of materials** – Embodied energy of materials used in constructing a building accounts for roughly 10 per cent of its total energy usage over 100 years. So although it is important, materials should be primarily chosen for their effect on the overall energy usage of the building.

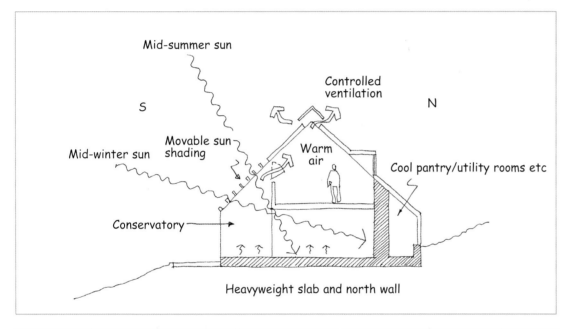

Mid-summer sun

Controlled ventilation

S

N

Mid-winter sun

Movable sun shading

Warm air

Cool pantry/utility rooms etc

Conservatory

Heavyweight slab and north wall

Above left **Good sustainable design focuses on how the whole building performs**

Left **This ecologically sound house supplies all its energy requirements by producing electricity from photo-voltaic cells on the roof**

Moisture and Ventilation

All houses contain moisture. This is generated by atmospheric conditions, breathing by occupants, steam from baths and cooking, and many other activities. The amount of moisture (water vapour) that air contains increases with temperature, until it reaches saturation (dew point). Not only does warm air contain more moisture than cold air, it also has a higher vapour pressure, which forces it out through the building envelope. Now the problems really begin, because when the warm, moist air hits a cold surface, it reaches its dew point and the moisture condenses on the cold surface. If this water condenses on a roof timber with no ventilation, for instance, rot can quickly set in.

Before the 1960s condensation within buildings wasn't much of a problem. Most houses had open fireplaces and plenty of natural ventilation drawing air constantly through the building. With changes in building techniques and the need to minimize heat loss (a third of a building's heat loss can be through bad draught-proofing), modern houses through necessity have become more hermetically sealed. To minimize the risk of condensation and provide constant fresh air but still maintain the airtightness of the envelope, ventilation has to be carefully designed. Building regulations stipulate that domestic housing should have constant background ventilation, which is normally supplied by trickle vents above windows or airbricks in the walls. Rapid ventilation is also required, and this is usually achieved simply by having windows that open. And finally, areas in the house where high moisture is generated, such as in bathrooms and above cookers, need to have extractor fans fitted. These can be of the 'passive stack' type, which work by air pressure drawing moist air up a tube through the roof, where it is vented out through a specially fitted ridge tile. Whole-house heat-recovery systems can also be used to ventilate every room in the building. (See mechanical ventilation and heat recovery in Chapter 8).

Above **Areas that produce a lot of moisture, such as this kitchen, should have extraction systems fitted**

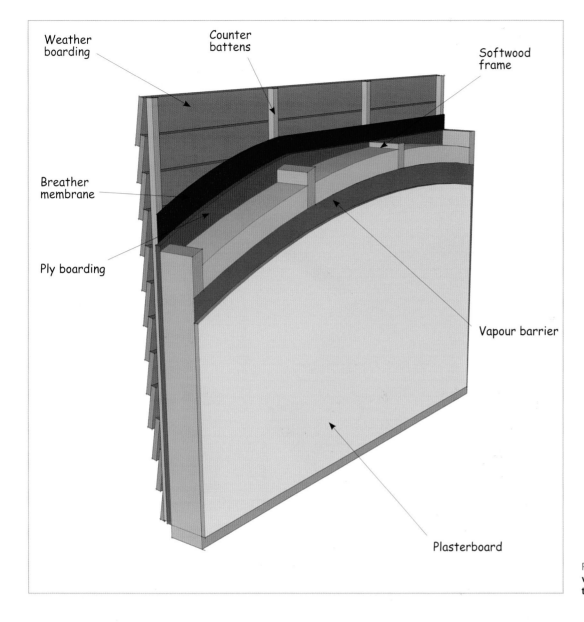

Weather boarding

Counter battens

Softwood frame

Breather membrane

Ply boarding

Vapour barrier

Plasterboard

Right **For a building to be able to 'breathe' properly, a vapour-control layer needs to be put on the warm side of the wall, and a breather membrane put on the cold side**

Breathing Walls

The building envelope needs to limit the amount of water vapour entering the building, whilst at the same time allowing a diffusion of water vapour out of it. This ability to 'breathe' is normally achieved by putting a vapour-control layer on the warm side and a 'breather' membrane on the cool side. To maintain a proper diffusion of water vapour through the building envelope, so that it doesn't condense within it (interstitial condensation), the vapour resistance of the vapour-control layer on the warm side should be at least five times greater than through the breather membrane on the cold side.

Walls constructed using a softwood frame usually have a reasonably vapour-resistant ply sheathing on the outside, so to correct this balance a vapour barrier needs to be placed on the inside surface. This would normally be a layer of polythene which is then covered in plasterboard. If a breathable bitumen softboard is used on the outside of the softwood frame instead and the cavity between the studs is filled with an insulation such as 'warmcell', then a material with a much lower vapour resistance is required on the inside. This can normally be achieved by using a plasterboard which is either foil backed or painted in a vapour-resistant covering. This type of breathing construction increases the exchange of water vapour which helps the internal environment from becoming either too dry or too humid.

Construction Systems for the Envelope

The materials used to enclose the main elements of an oak frame, such as the roof, walls and floor, can be distinguished by their mass. The systems made up by these materials can be divided into three categories: lightweight, heavyweight and hybrid (using a combination of both). Using the right system can significantly improve the thermal and acoustic performance of a building and at the same time have a minimal environmental impact. Ultimately the right choice will depend on a number of design and construction issues, such as the building's orientation, the use of passive solar heating and the level of occupancy. In most cases, the hybrid approach of using a combination of materials will provide the best overall solution. The exterior appearance of the envelope can look the same for either lightweight or heavyweight systems. For instance, a cavity block or a timber stud wall can both be faced with weatherboarding or alternatively with brick.

Below **An oak frame that has been enclosed with natural stone to blend in with local environment**

Below left **The weatherboarding on this house was put onto a high-density cavity-block wall, which encloses the oak frame**

Lightweight Construction

Lightweight materials used to construct the envelope include softwood stud frames and structurally insulated panels (SIPs). Lightweight buildings generally have the following properties:

∧ **Fast response** – They have a fast response to heating, so they warm up quickly. This is obviously an advantage if the building is not continually occupied or heated.

∧ **Airtight** – They are easier to make airtight, which is one of the main ways of conserving energy.

∧ **Low energy** – They usually have a very low embodied energy.

∧ **Heat loss** – They are not as good at storing heat, so passive solar heating is not as effective.

∧ **Acoustically poor** – They don't offer very good sound resistance, as mass is the most effective acoustic insulation. Acoustic properties can be improved if additional layers of sound-deadening boards are used within the construction, although this is likely to be costly on materials.

∧ **Maximized floor space** – Where space is at a premium, timber-stud walls and SIPs are typically much thinner than masonry walls, so more floor area can be achieved for the same overall footprint. Unfortunately window and door reveals are also thinner, which some people find aesthetically displeasing.

∧ **Fast drying** – They can be constructed with 'dry' materials, which means the building won't take months to dry out after the construction has finished.

Softwood frames are typically made out of 5½x1½in (140x40mm) treated-softwood studs, which are then covered on the outside with a sheathing board. They can be made up on-site or bought in as ready-made panels from specialist companies. A variety of insulation products can be used to infill between the studs (see Insulation on page 161). A vapour-check barrier needs to be put on the inside of the stud wall and a breather membrane on the outside. A ventilated cavity needs to be formed on the outside of the breather membrane, after which a variety of materials can be used as cladding. Before fixing the softwood frame, it is a good idea to fix a plywood packer (the same thickness as the finishing plasterboard) on the outside of the oak frame, about 1in (25mm) in from the edge. Once the first fix on the plumbing and wiring is complete, the internal plasterboard finish can then be slid between the oak and softwood frames and fixed into position. Unsightly gaps should then be avoided between the plasterboard and the frame when the oak shrinks.

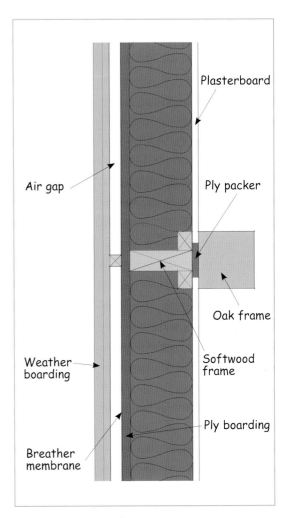

Air gap

Plasterboard

Ply packer

Oak frame

Softwood frame

Ply boarding

Weather boarding

Breather membrane

Above right A plywood packer should be fitted to the outside of the oak frame so that the plasterboard can be slid behind the oak

Right A softwood frame can be fitted to the outside of the oak and the services can be run in prior to the insulation being fitted

SIPs are constructed by sandwiching a highly insulated core between sheathing board (normally OSB). Structurally they are very strong and rigid. They also provide an almost complete layer of insulation, with no cold bridging, so perform well thermally and can easily be made to be air-tight. They can be cut on-site to fit a frame, although this is normally done off-site by the manufacturer, working from the architectural and framing drawings. They are quick to apply to both the walls and roof of an oak frame, and thus can produce a water-tight building in just a few weeks. This means that the frame can be enclosed quickly, cutting down the site time and allowing the follow-on trades faster access. They cannot be applied directly to the oak frame, because, as with softwood frames, a gap should be left for the internal plasterboard. Unlike softwood frames, though, a service gap also needs to be provided for the first-fix plumbing and wiring. This is normally achieved by first fixing a 2in (50mm) thick counter batten all over the outside of the frame. The SIP is then attached to that using long screws. A smaller batten is then fixed to the inside of the SIP to create a service void. A continuous vapour barrier should be attached to the inside of the SIP and a breather membrane to the outside.

Above **Lightweight materials, like these SIPs, can be used to enclose a building quickly**

Top right **A masonry wall under construction**

Bottom right **If designed correctly, heavyweight structures can store and retain heat received from passive solar systems**

Left **SIPs should be fixed to counter battens on the frame, so a gap is created between them and the oak. Services such as plumbing and wiring can then be run before the plasterboard is finally fitted**

Exterior cladding

Plasterboard

Softwood spacers

Oak post

Services

Insulation

OSB

OSB or Oriented-Strand Board evolved from waferboard in the late 1970s. Long wood strands are orientated, as opposed to being randomly placed, and it is used for similar functions as plywood of a similar thickness.

Heavyweight Construction

Heavyweight construction materials typically include concrete, concrete blocks, stone and bricks. Surprisingly, rendered 'straw bale and water' can also offer high-mass solutions. Heavyweight buildings generally have the following properties:

∧ **Constant temperature** – They are less prone to temperature fluctuations but take much longer to heat up. They work more efficiently if the temperature within them is kept constant; in other words they are designed with constant occupancy in mind.

∧ **Efficient heat distribution** – They remain cooler during the summer because the fabric of the building takes a long time to heat up during the day when it's hot. The stored heat is then released slowly during the night when it's cooler. They also have the ability to store and emit heat gained from properly designed passive solar systems during the winter.

∧ **Good sound insulation** – The thickness and density of heavyweight construction means that acoustically they perform very well, and should be considered for buildings in noisy environments.

∧ **High embodied energy** – They use materials with high embodied energy.

Heavyweight materials can make use of solar energy by storing the heat gained during the day and releasing it slowly during the night. The use of solar energy does require the building to be orientated correctly. Cavity-block construction is the most common form of heavyweight walling currently used in the UK. To make use of thermal mass, insulation needs to be put on the outer face of the inner block, within the cavity. A cavity of 2in (50mm) should be maintained so the gap between the inner and outer skins can become large, so extra long cavity ties have to be used which can be expensive. Construction of block walls is obviously all on-site, which needs skilled tradesmen and can also be prone to delays from bad weather. Unfortunately concrete blocks are not particularly 'environmentally friendly' per se, but as mentioned previously, the embodied energy of a material has to be considered within the total lifetime energy usage of a building.

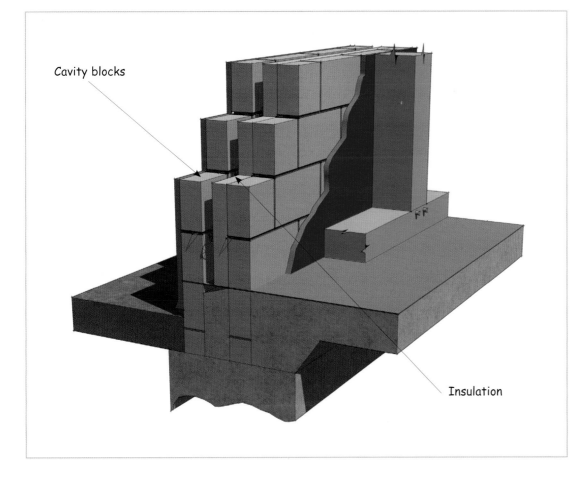

Cavity blocks

Insulation

Heat isn't the only thing that insulation can control. Noise is also a problem for many people in different locations. The exact amount of noise that your building is subject to will obviously depend a great deal on its location and orientation, and appropriate acoustic insulation can make a really big difference to your ultimate quality of living.

Left **Properly designed high thermal mass buildings can store heat received from the sun during the day and release it slowly at night when it is cooler**

Below **Acoustic insulation was placed above this roof to reduce the noise from the external environment**

Hybrid Walls

A hybrid walling system combines both lightweight and heavyweight materials in an attempt to achieve the best of both worlds. Thin, lightweight walls which contain large amounts of insulation can be used for most of the construction. Fast construction time and the ease of making them airtight are obviously an attraction. Heavyweight materials such as concrete block walls and concrete floors should be positioned to obtain the best use of any passive solar gains. These do not want to be covered up with insulating materials, such as carpets, which will stop them warming up. Floor tiles and painted dark surfaces should be used instead, which will help absorb the heat.

The Roof

The best place to start finishing off the frame is on the roof. With most traditional forms of construction the walls are slowly built up until they reach wallplate level and then the roof structure is made and eventually covered. With an oak frame the roof structure is erected with the rest of the frame, so covering it can begin immediately. There are many advantages in making the roof watertight as quickly as possible, especially considering the capricious nature of the weather. Once the roof is on, the inside of the frame can become a makeshift workshop to help keep builders and materials dry, during the rest of the construction. It will also slow down the drying process of the oak used in constructing the frame. This must be a very important consideration when constructing with green timber, because the slower it dries, the less shrinkage will occur. Even in middle of the winter the frame will dry out quicker if it is not covered, but in the summer the process is greatly speeded up.

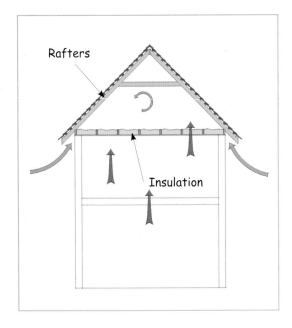

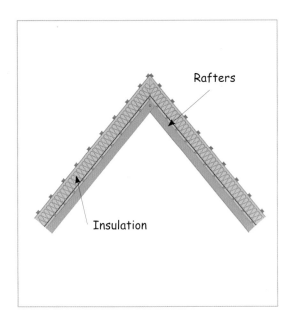

Warm and Cold Roofs

Roofs are an obvious target for condensation as the warm moist air rises upwards and escapes onto the cold roof outside any insulation. Traditionally most houses were built with a loft that was meant to be cold – in other words the insulation was put above the ceiling of the upstairs rooms and the space above it was unheated and ventilated. Moisture escaping into the roof space was well ventilated by the movement of air and therefore caused no problems. This type of roof is naturally called a cold roof. Typically, oak-framed houses do not have a conventional loft space but tend to be designed so that they are open right up to the rafters.

It is not suitable to place roof insulation between oak rafters, because generally they won't be deep enough and it would be a shame to cover them up. This leaves one option, and that is to put it above the rafters to create a warm roof. If all the structural timbers are on the warm side of the roof, their temperature should always be above the dew point and therefore not subject to moist air condensing on them. This also means that the roof space above does not need ventilation, unlike cold roof construction. As with walls, a vapour barrier needs to be placed on the warm side of the insulation and a breather membrane on the cold side.

Sarking on Top of Oak Rafters

If oak rafters are going to be left exposed, a surface that can be decorated or one that doesn't need any further decoration needs to be applied above them. The two usual choices are either plasterboard or wooden sarking. The effect of shrinkage in the oak rafters means that over time more of the covering material will be exposed than first seen. This won't matter if you decide to use a natural timber board as your sarking material, as you will just see a tiny bit more of it. Unfortunately though, if you use white plasterboard and then decorate it after laying it with a dark paint, then thin white lines will appear down every rafter as they shrink. The solution to this problem is either to use materials that don't need further decoration, or to decorate the materials **before** they are laid. This is easier with sarking board as it can be simply painted and left to dry on the ground before being nailed over the rafters. Even if it gets slightly damaged in transit, it can soon be touched up from below. It is best to use a board that is tongued and grooved, so if there is some shrinkage in the board, chinks of light are not seen reflecting off the shiny surfaces from the insulation. If tongue and groove boards are not available, dark building paper can be laid over the sarking to screen any silvery surfaces. A variety of timbers can be used for

Top **Traditional cold-roof construction has insulated ceilings and cold vented roof spaces**

Above **In warm-roof construction the insulation is placed above the rafters. A vapour barrier should be placed on the warm side and a breather membrane on the cold side**

Right **Plasterboard sarking is applied on top of oak rafters to counter the effects of shrinkage**

Above **Sarking can be painted different colours to emphasize a change of living zones within the house**

Left **Sarking should be painted before it is laid on the rafters; otherwise, when the rafters shrink, two thin stripes of undecorated boarding will become visible**

Sarking

Sarking boards are close-set boards that are designed to carry the roofing material and are used as lining.

sarking, from cheap softwood floorboards that can be painted, to expensive kiln-dried oak that needs no further treatment. Sheet materials like plasterboard are harder to lay without damaging them. Another problem with plasterboard is the joints created between them when they are laid. They need to be filled either by plastering the whole board (not a good idea because of shrinkage again) or dry-lined (just the joints are taped and plastered). If possible, the boards should be cut so any joints are hidden either over a purlin or a rafter.

Insulation

There is a wide choice of insulation products available in today's market, added to which new and better materials are being developed all the time. As the building regulations increasingly highlight the need to produce more thermally efficient buildings, manufacturers of insulation products have responded by producing thinner materials which achieve the same U values as the older, thicker ones. Much of this drive towards thinner materials has to do with meeting the modern requirements, whilst

Left **A vapour barrier should be placed on top of the sarking before any insulation is laid**

Below **Sarking can be lightly stained so that the wood is still visible**

still maintaining conventional building techniques. For instance, a conventional timber-stud-frame wall is built with 3½in (90mm) deep studs, and so can only hold 3½in (90mm) of mineral insulation. If this no longer passes the required U value, either a better insulation needs to be used or the stud has to be increased to 5½in (140mm) to hold more. What is actually important in choosing an insulation material is the overall thermal design of the building's envelope, as all insulation materials will achieve the same U value at varying thicknesses. The criteria for choosing which insulation material is used will ultimately hinge on how the external envelope is constructed. Ideally it needs to have a long life to maximize its energy-saving potential, it should be produced out of a zero-ozone-depleting material and if the envelope thickness needs to be restricted, the best thermal insulator should be used. What follows is a summary of the properties of the most common insulation materials available today.

∧ **Mineral wools** – These are made out of crushed minerals which are then heated and blown into fibrous material. Products made this way include fibreglass and rockwool. They are cheap but need a thick layer to produce a good U value, which needs to be considered when designing the building's envelope. They are normally sold in rolls and have traditionally been used in lofts, but can sag if placed in walls and pitched roofs. For use in walls, an extra manufacturing process is applied to produce stiffer 'batts', although they can cost nearly double. If you have ever tried laying mineral-wool insulation, you will know that it can cause severe irritation of the skin and eyes, and can also produce harmful airborne dust, so protective equipment should be worn when handling them.

∧ **Oil-derived products** – Most products that fall within this category are usually sold as some type of rigid board. In the past, polyurethane was commonly produced using CFC-ozone-depleting potential (ODP) blowing agents. This has now been phased out and replaced by other blowing agents. Some, like HCFCs, still have some ODP but others such as Pentane and CO_2 have zero ODP. The market is dominated by Kingspan and Celotex, who produce a range of foil-faced sheet-sarking boards, which have good thermal-resistance properties. If space is an issue, then they are a good choice as some mineral wools require twice as much depth to achieve the same U value.

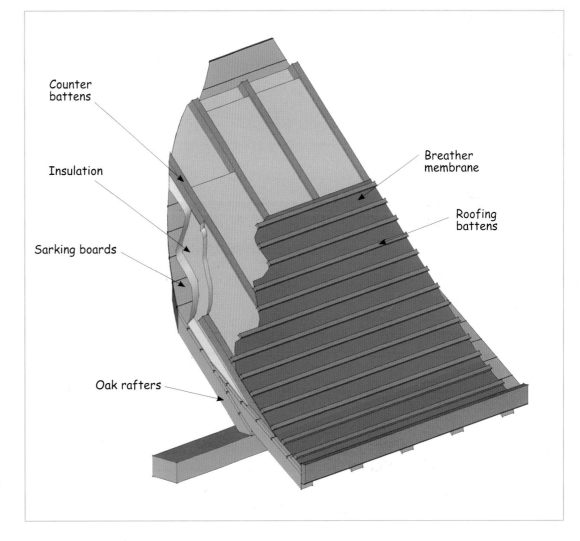

Left **A well insulated home will stay warm during the winter and be cool in the summer**

Right **Rigid-board insulation is a good option if space is an issue. In roofs normally two layers are needed: one complete layer over the sarking to prevent cold bridging, and one between the counter-battens to add to the insulation. A vapour barrier needs to be added to the warm side and a breather membrane to the cold side**

Counter battens

Insulation

Sarking boards

Oak rafters

Breather membrane

Roofing battens

∧ **Cellulose fibre** – Known as 'warmcell', it is made from recycled newspaper, and is a very 'green' insulation product. It is supplied as a dry-loss fibrous material that needs to be specially blown into the buildings envelope. It works well with a softwood frame, where it is blown in from holes cut in the outside sheathing at high pressure after the internal plasterboard has been fixed. The advantage of this method of application is that, theoretically, every tiny nook and cranny within the wall can be filled, even around pipes and wires. It is also available in a manufactured structurally insulated panel (SIP), made out of two outer layers of sheathing board encasing composite beams and the insulation.

∧ **Sheep's wool** – No surprises that sheep's wool makes a good insulator, as we have been wearing it for centuries! It's manufactured by mechanically bonding new sheep's wool that is washed and treated with a pest repellent. It is supplied in a roll and used in the same way as the mineral wools, except it's not irritating to handle.

∧ **Straw bale** – Straw bales can be used to build the walling envelope of an oak framed building. They act as an excellent insulator on their own, mainly because of their thickness, which typically is 18in (450mm). Some tests have reported U values as low as 0.13 when using wheat bales laid flat. This value is much better than currently required by building regulations. When rendered on both sides with thick coats of lime mortar they can also add thermal mass to a building.

Above **These SIPs are made with polyurethane injected between two layers of OSB board**

Bottom left **Mineral wool laid between multi-web joists**

Bottom right **Straw bales are an excellent insulator and they can also add thermal mass to a building if rendered with lime mortar**

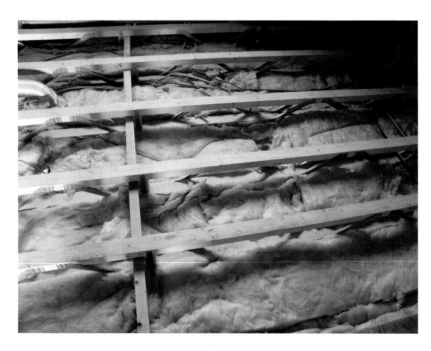

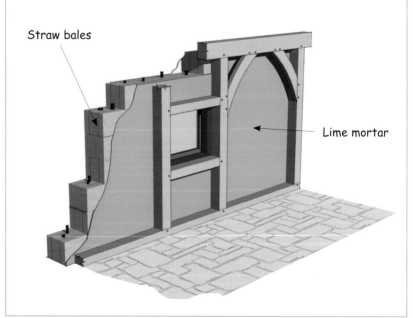

Straw bales

Lime mortar

The Building Envelope

⋀ **The outer skin** – The building envelope should ideally achieve good 'U values' and airtightness. Thermal insulation materials should be applied in complete layers so they are more effective. Timbers exposed both internally and externally can act as a 'cold bridge' and can also allow air through cracks as the frame dries.

⋀ **Sustainability** – This is a major issue with any new building, and, while oak-framed buildings are relatively environmentally friendly, factors such as orientation, minimization of heat loss, passive solar design, the use of energy-efficient appliances, proper ventilation and the choice of the right materials will all help to minimize the impact of the building on the environment.

⋀ **Ventilation** – This is an important consideration with any house but especially with an oak-framed one because of the risk of rot setting in. To minimize this risk ventilation has to be carefully designed. Building regulations ensure background ventilation, while rapid ventilation is left to windows that open. Bathrooms and kitchens should also have extractor fans fitted. Mechanical Ventilation and Heat Recovery systems are more efficient than traditional forms of ventilation. 'Breathing walls' limit the amount of water vapour entering the building, but allow it to diffuse out.

⋀ **Construction systems** – These are normally categorized, depending on the mass of the materials used, as lightweight, heavyweight or hybrid. Lightweight buildings will warm up and cool down quickly and don't offer very good sound resistance. However, the relatively thin walls allow more floor space for a building of a particular size. Heavyweight construction uses materials such as concrete, stone and bricks, and even rendered 'straw bale and water'. They generally suffer from fewer temperature fluctuations but take longer to heat up; because of this they are more efficient if the temperature is consistent – if occupancy is constant. Hybrid walls contain both lightweight and heavyweight materials: lightweight walls containing a lot of insulation are used for most of the construction, while heavyweight materials are positioned to make the most of any passive solar gains.

⋀ **The roof** – The roof is the best place to start finishing the building as it can protect the work that is being carried on underneath it from the elements. It also helps to ensure that the frame will dry out slowly and therefore less shrinkage will occur. Most oak-framed houses are open to the rafters with the insulation above, which is called a warm roof.

⋀ **Insulation** – There are many insulation products, both modern and traditional, available to help in the drive to produce more thermally efficient buildings. These include: mineral wools, oil-derived products, multi-foiled quilts, cellulose fibre, sheep's wool and straw bale.

Sustainability at the Gateway Centre

The Gateway Centre, completed in 2004 at Cotswold Water Park, UK, was opened to the public by the well-known conservationist David Bellamy. The visitor centre, which is located on the side of a beautiful lake and wildlife reserve, contains a café, exhibition centre and shop. From the onset it was designed with sustainability in mind, which was based around the green-oak frame.

Top **Photovoltaic (PV) cells on the roof of the Gateway Centre generate electricity for the building and power the ground-source heat pumps**

Far left **South-facing glazing provides passive solar heating. The glass is protected from over heating by the summer sun by a large overhang on the roof**

Left **The Gateway Centre contains a café, exhibition centre and shop**

The frame was an aisled construction containing nearly 40 tons of oak. It was all traditionally made out of certified timber, from a sustainably managed woodland. The exterior envelope was hung off the outside of the frame which wrapped it in a complete layer of insulation. This prevented any 'cold bridging' through the oak. The walls were constructed with a softwood insulated frame and covered with locally produced weatherboarding. The majority of the glazing is located on the south side of the building overlooking the lake. The glazing provides the building with heat gained by the passive solar effect of sun shining through the double-glazed units. The glass is protected by a large overhang to restrict the summer sunshine whilst permitting the winter sunshine to enter the building when the sun's orbit is lower.

The building is open to the rafters and has a warm-roof construction with the insulation placed above the rafters. On top of the roof is a bank of 300 photovoltaic (PV) panels that generate up to 25 kilowatts of electricity an hour on a sunny day. This is enough electricity to run nearly four and a half houses. The electricity is used to power the visitors' centre and run the climate-control pumps. Any surplus electricity is sold back to the national grid.

The temperature within the building is controlled by energy collected from the lake using a ground-source heat pump. Almost 3½ miles (5.5km) of plastic pipe was sunk around the lake in a closed loop containing water (with some antifreeze in it). This is connected to the building and continually pumped around it and the lake using super-efficient heat pumps, powered by electricity generated from the PV cells on the roof. The stored energy from the lake water is transferred by the liquid in the pipes into radiator-type units located around the building. These units then transform the energy into either hot or cold air depending on the ambient temperature. Typically, the system will produce three or four times as much thermal energy as is used in electricity to drive the pumps. The rainwater collected on the roof is not wasted either. This harvested water is stored in a big tank under the ground and then is re-used to flush toilets, thereby saving valuable drinking water. All in all, this is a truly sustainable building.

Below **In a warm-roof construction the insulation is placed above the rafters in a continual layer to prevent cold bridging from occurring**

Bottom right **The building is heated by a ground-source heat pump, which transfers energy from the lake to the building via 3½ miles (5.5km) of liquid-filled pipe, sunk at the bottom of the lake**

Chapter 8

Finishing

Finishing off any building requires care and attention, but this is particularly important if it is an oak-framed building. Green oak will shrink, so allowances must be made for this by careful detailing and design during the finishing process. This chapter examines how glazing and flooring interact with an oak frame.

Glazing Frames

Medieval oak-framed houses were dark and gloomy places, lit only by a few small windows and the fires and lamps that burnt inside. Up until the late sixteenth century, most windows didn't even have glass, and shutters were used to keep the weather out. Today, the technology of glazing has enabled us to design houses that are flooded with natural light. Unfortunately, glass does not have good thermal-resistance properties, so the amount that is used, and where it is positioned within a building, needs to be carefully managed. The optimum area of glazing to use strikes a balance between the heat lost through it and the energy saved from using daylight. This is typically between 20 per cent and 30 per cent of the total floor area, although this could be increased if insulated shutters and shading are used.

Left **Oak-framed barn-style house**

Above **Glass is not a very good thermal insulator so glazing needs to be carefully designed**

Left **An example of medieval glazing**

Below left **Light flooding into a modern oak-framed house**

Right **Passive solar design should be considered when glazing large areas. If too much glass is used, overheating can occur on sunny days whilst heat can be lost from the building on cloudy days**

Oak-framed buildings can be designed so that the framing members define and enhance any glazed areas. This needs to be considered at the initial design stage, so the best use of the sunlight within the plot can be achieved. Large areas of glass are conventionally concentrated in a conservatory, which is connected to the house, but usually isolated by means of a closable partition or doorway. The trouble with conservatories is that they have a tendency to overheat during warm, sunny spells but don't have the ability to retain any heat when it's not warm and sunny. If the glass on the roof is replaced with insulation and tiles, more of the heat can be retained and suddenly the space starts to work more efficiently. If the partition between it and the house is also removed, it now can become a room that can be used all year round. The advantages of having a glazed area is the feeling of light and space it produces, bringing the outside environment into the building.

Passive Solar Heating

Glazing large areas is a balancing act between heat gain and heat loss. If too much glass is used, the interior can overheat on sunny days but lose more heat on cloudy days, resulting in a net loss. But if designed correctly, an area of glazing positioned on the south side of a house can bring passive solar heating gains, which helps to improve the thermal efficiency of the building. This works by retaining the heat which is naturally produced from sunlight. The sun emits heat radiation at short wavelengths, which passes through the glass easily, and this in turn warms the internal surfaces of the building. The building then radiates the heat back at much longer wavelengths, which do not pass so readily through the glass, so consequently the heat is trapped, warming the interior.

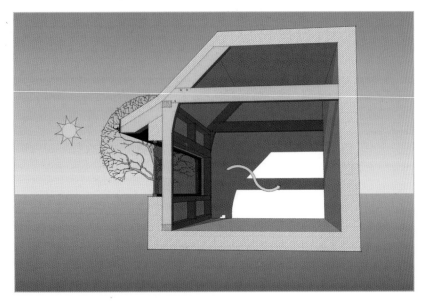

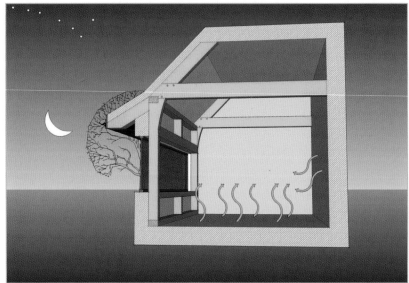

Passive solar heating systems need to be designed to work with areas of high thermal mass – usually masonry. The heat generated during the day by the sun is trapped within it and then re-released slowly during the night when the temperature is lower. High thermal mass can be achieved by having masonry walls within the building. These need to be insulated towards the outside of the building so that the stored heat isn't lost. Another way of achieving high thermal mass is to use a solid concrete floor in front of the glazing. This shouldn't be covered with a carpet that will insulate the floor from being heated. The windows on the north side should also be restricted in numbers and size, to limit the loss of heat by that route. Further improvements can be made by shading the glazing with large roof overhangs. This prevents overheating by the summer sun when it's high in the sky, whilst permitting the winter sun to enter when its orbit is lower in the sky.

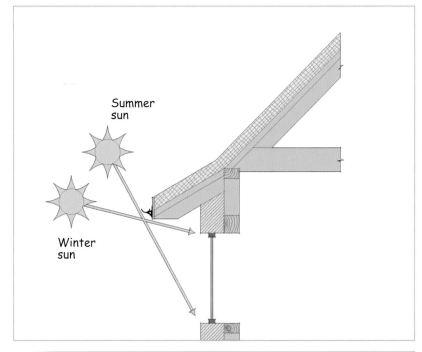

Summer sun

Winter sun

Above **Heat generated from sunlight during the day is absorbed by high-mass materials**

Above right **Absorbed heat is slowly released over the course of the night**

Middle right **Overhangs on roofs should be designed so that they restrict summer sun but allow winter sun to shine through**

Right **Overhanging roof on a contemporary frame**

Opposite page top **Heat from the sun is produced in short wavelengths, and passes easily through glass. Heat reflects back from the interior in longer wavelengths, and does not pass so readily back through the glass**

Opposite page right **Sunlight flooding in on a winter's day**

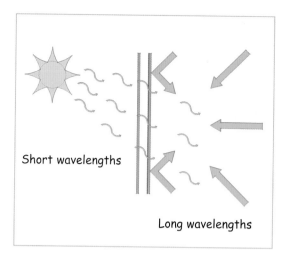

Short wavelengths

Long wavelengths

Glass

Glass is obviously going to make up most of the area in any glazing system, so the choice of what type to use is going to be crucial to the overall success of it. Double-glazed units with toughened glass are an absolute minimum requirement both for safety and thermal efficiency. It is advisable, though, to use a unit that will provide the best U value possible, especially as the cost of the glass will probably make up only a small proportion of the total cost to direct-glaze a frame. In fact, a glazed area will probably not even pass building regulations unless double-glazed units with a low U value (see page 148) are used. To improve the efficiency of a unit, the internal surface of glass that faces the air gap can have a low-emissivity coating which helps stop heat radiating out of the glass, reflecting it instead back into the building. A further improvement is made if an Argon-gas-filled unit is used in conjunction with a low-emissivity coating.

Emissivity

Emissivity is a complex scientific term to do with the capacity of certain molecules to emit and absorb energy. In lay terms it is the comparison of a selective emitter with a perfect emitter at the same temperature. The result is a fractional indicator of how much energy is emitted from a particular material at a particular temperature compared with a theoretical value for a perfect emitter at the same temperature. The lower the fraction the less energy is radiated and therefore the more efficient an insulator the material is.

Direct Glazing

An oak-framed building can either be glazed by directly fixing the glass onto the structural frame, or alternatively the glazing can form part of the external envelope, as is the case with normal windows. Large window units can be specially designed to fit whole bays of the frame and, because they are fixed into the external envelope and not the green-oak frame, movement should not be a problem. With direct glazing, on the other hand, the glass is in contact with the green-oak frame, and some movement caused by shrinkage is going to be inevitable. The basic principle of direct glazing is to place the glass onto the outside surface of the green-oak frame. This is then held in place by a stable air-dried-oak cover piece, creating an oak 'sandwich', with the glass as the 'filling'. The stable dry oak is on the outside of the building, maintaining the weather barrier, whilst any potential movement from the green oak will occur on the inside, and therefore will not affect the performance of the glazing. Although this system has been designed to cope with green oak, excessive shrinkage or movement could potentially compromise it. To minimize this risk, areas of the frame which are going to be glazed should be made out of clean and stable oak – in other words, oak that has been air-dried for a while and has no structural defects that could cause it to twist or deform.

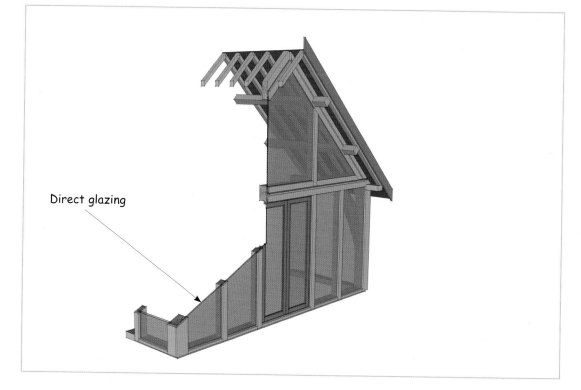

Left **Glazing fixed on to the external envelope**

Above right **Direct glazing fixes onto the outside of the structural frame and is attached with dry-oak cover piece**

Right **A section through direct glazing**

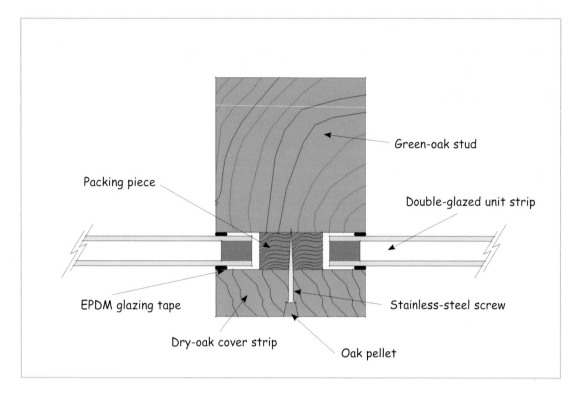

Packing piece

Green-oak stud

Double-glazed unit strip

EPDM glazing tape

Stainless-steel screw

Dry-oak cover strip

Oak pellet

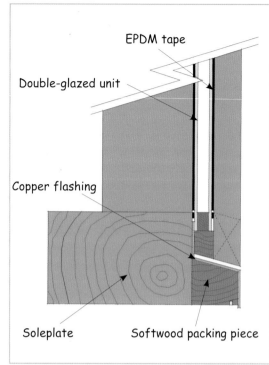

EPDM tape

Double-glazed unit

Copper flashing

Soleplate

Softwood packing piece

A typical double-glazed unit will be 1in (25mm) thick, made up of two layers of glass ⁵⁄₃₂in (4mm) thick with a ⁵⁄₈in (16mm) gap. The overall size of the unit is governed by how much it overlaps the oak frame that is supporting it – typically this is 1⅛in (30mm) in every direction, which is enough to support it and cover the spacing bar. The glazed units are separated by a softwood packing piece that is 1⅛in (30mm) thick, which is fixed into the vertical oak studs. The outside edge of the stud then has an EPDM glazing tape stuck to it, which is a type of rubber with a closed-cell structure very much like the gaskets used to seal car doors. The tape should be ³⁄₁₆in (5mm) thick and on one side have a self-adhesive glue on it. This is also applied to the inside edge of the air-dried cover strip so when it is screwed into the packer, the EPDM tape compresses to ⅛in (3mm) thick on the double-glazed

unit and clamps it into position. This is a dry-glaze system, which means that the glass is not bedded in any mastics or silicone sealant, so it is important to leave a gap of at least ¼in (6mm) between the edge of the unit and the softwood packer. If not enough of a gap is left, any water that does leak into the glass could be sucked into the inside of the frame by capillary action, instead of draining harmlessly out. The oak cover strips need to be air-dried to a moisture content of between 15 per cent and 18 per cent so they will be as stable as possible in the outside environment. The timber also needs to be good joinery-grade oak that is unlikely to cup or twist. It should be fixed into the packing pieces with stainless-steel screws, into countersunk holes that are later filled with oak pellets. This enables cover strips to be removed (although with some difficulty) if ever a double-glazed unit needs to be replaced.

The glazing detail on the vertical studs is quite simple, but on horizontal members the situation becomes much more complicated, especially if they are dividing an upper and a lower section of glazing. It becomes necessary to introduce a metal flashing at the bottom of the upper units to stop any water ingress. The flashing is tucked in behind the outer pane of glass, which is specially extended past the spacer bar, and then formed over the header of the unit below. The best material to use for the flashing is ¹⁄₃₂in (0.6mm) copper, which is relatively easy to bend (or can come pre-bent from a supplier) and doesn't react with the oak. Lead can also be used but should be coated with a bituminous tar, otherwise the tannin in the oak is likely to eat it away.

If opening doors and windows are required alongside the fixed direct glazing, they should always be hung in their own stable frames. This means that whatever movement there is within the structural frame, the action of the doors and windows won't be affected. The frames can be fixed by the dry-oak capping pieces in the same way as a double-glazed unit, if they are made with an extra protrusion on the front edge. This needs to be the same thickness as the double-glazed unit – 1in (25mm) – and overlap the frame by the same amount of roughly 1⅛in (30mm). It can then be treated the same as the glass and held in place by the cover strips and glazing tape. An

EPDM

EPDM (Ethylene Propylene Diene Monomer) has been in use on roofs since the 1960s and is relatively inexpensive. It is also easy to use and has minimal odour, a good level of colour consistency and can be used in either its vulcanized or non-vulcanized form.

Bituminous Tar

The term bitumen is used to denote different materials depending on where you live in the English-speaking world. In this context it is used specifically to mean a black viscous tar that is formed of organic materials.

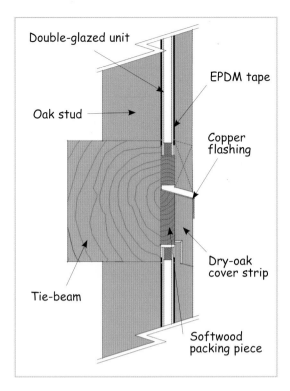

Double-glazed unit

EPDM tape

Oak stud

Copper flashing

Dry-oak cover strip

Tie-beam

Softwood packing piece

extra bead of wood is placed against the inside of the window or door frame to cover up any gaps left by shrinkage in the structural frame. The bead should be fixed into the structural oak only, which will allow the window or door frames to slide independently between it and the cover strip if there is any shrinkage.

Far left **A section through a stud. A softwood counter-batten is fixed onto the structural frame, and the glass is clamped in place by a dry-oak cover piece**

Above left **A side section looking through a soleplate**

Above **A side section through a tie beam. Copper or lead flashing needs to be placed between the top and lower glazing units**

Top right **Direct-glazed oak-framed extension**

Right **Doors and windows can be fixed to the frame in the same way as the glazing units**

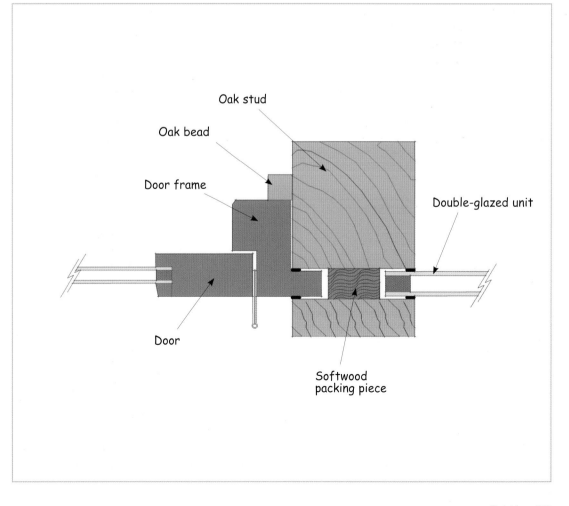

Oak stud

Oak bead

Door frame

Double-glazed unit

Door

Softwood packing piece

Left **Roof glazing should be restricted to a minimum**

Below **Roofs can be glazed by a modified direct-glazing system but laminated glass should be used for safety**

Inset below **An enlargement of a modified direct-glazing system showing the laminated glass attached to the bottom batten**

Bottom right **Glass on a roof should be made stronger than on the walls**

Glazing Roofs

Too much glass on roofs is never a good idea. It causes the inside to overheat in the sunshine and, because hot air rises, lets it all out again when it's cloudy. Added to that, it's difficult to clean and maintain safely – you can see why I'm not keen on it. If only a small area of roof glazing is required, it's best to use one of the many manufactured roof lights available today. For a larger area, though, it is possible to glaze a green-oak roof, in exactly the same way as the direct-glazing method described above, but with a few small modifications. The glass should be made stronger than on the walls, so both the outer and inner pane should be at least ¼in (6mm) thick. The outer pane, like the walls, should be made of toughened glass and the inner pane should be made of laminated glass. If anything falls onto the roof and the outer pane breaks, the laminated glass shouldn't shatter, and will hopefully protect anyone below. The glass needs to be made up into a stepped double-glazed unit, with the outer leaf extending far enough to reach any guttering, and the bottom spacer bar resting against a restraining piece of wood to stop the glass slowly creeping down the roof. The rest of the glazing system is made up in the same way as the walls, except that the glass will be resting on the sloping rafters instead of the vertical posts. If the glazing doesn't extend to the top of the roof, a trimmer needs to be inserted between the rafters to take the head of the unit. A copper or lead flashing can then be dressed over the top cover piece and under the tiles above, to stop any ingress into the roof of water or wind.

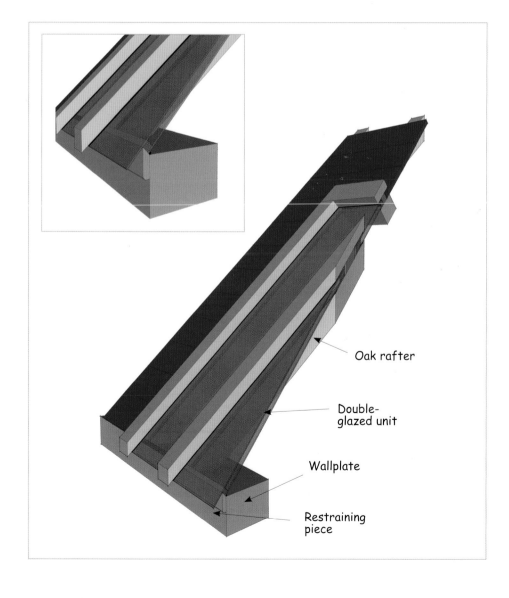

Oak rafter

Double-glazed unit

Wallplate

Restraining piece

Floors

A variety of flooring systems can be used within an oak-framed house. Typically the design of the ground floor (normally built off the floor slab) will be different to that on the first floor (normally built on timber joists). The effects of shrinkage within the frame again need careful attention, but as long as this is allowed for it should cause no problems. The flooring zone, whether it be on the ground or upper floor, should be designed to fulfil several different functions.

∧ **Thermal insulation** – Each floor should contain insulation to stop heat escaping.

∧ **Acoustic insulation** – Floors should be designed to limit the passage of sound.

∧ **Service zone** – Floors should have an area in which ducting, pipes and cables can be run.

∧ **Underfloor heating** – If used, needs to be allowed for in the floor build-up.

∧ **Thermal mass** – Correctly designed high-thermal mass floors will help save energy.

Left **Oak flooring with underfloor heating below**

Right top **Underfloor-heating pipes on top of insulation, with a screed covering and a wooden floor stuck on top**

Far right **Underfloor-heating pipes laid above oak joists**

Right bottom **Underfloor-heating pipes**

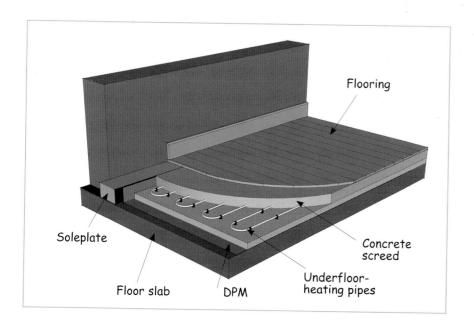

Flooring

Soleplate

Floor slab   DPM   Underfloor-heating pipes   Concrete screed

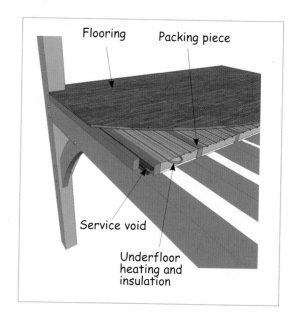

Flooring   Packing piece

Service void

Underfloor heating and insulation

Underfloor Heating

Underfloor heating was first used in Roman times. Recently it has seen a resurgence in popularity, and works especially well within oak-framed buildings. Most systems work by carrying warm water around a series of pipes under the floor. An underfloor heating system works at a much lower temperature than a conventional radiator set-up but feels more comfortable to live in, because it generally covers a large area and transfers heat at a temperature only marginally higher than the rooms above it.

Ground Floor

On the ground floor the underfloor heating pipes are normally laid on top of the structural slab or block-and-beam floor, between the soleplates. They need a layer of insulation below to make sure the heat flows into the room and not the slab. The pipes are connected to separate manifolds for each zone (and sometimes need to be pressure tested) and then covered in a layer of screed. The screed needs to be completely dry before the underfloor heating can be turned on. At this stage the final floor covering can be laid directly onto the screed.

Upper Floor

When laying a floor onto oak joists, all the materials used in the floor make-up, including the finished ceiling surface, should be put on top of the oak. In this way, when the joists shrink unsightly gaps shouldn't appear. The first layer to be applied should be the finished ceiling surface you want to see between the joists. It is basically the same as putting sarking board on the roof, so pre-decorating

Screed

Screed is a mixture of cement, sand and water that is applied to a concrete slab to give it a smooth finish on which the final flooring surface can be laid.

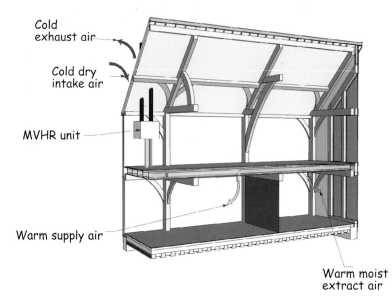

Cold
exhaust air

Cold dry
intake air

MVHR unit

Warm supply air

Warm moist
extract air

is advisable. If plasterboard is used it should be fixed between packing pieces applied to the middle of the joists. The weight of the floor above can then be transferred onto these rather than onto the plasterboard, which could cause it to crush. If timber is used for the sarking material a layer of plasterboard should be applied above it to provide fire and sound resistance between the floors. Current building regulations should be checked and advice sought from the underfloor heating supplier for the exact details. Above the sarking, insulation and then underfloor heating pipes are laid, with a void provided for a service zone. This can either run around the edge of the room within the insulation layer, or counter-battens can be laid on top of the insulation to make an area above it. The finished floor covering can now be fixed. This would normally be either chipboard, on which carpet or floor tiles can be laid, or timber floorboards, which again should be stored at the correct moisture content and left to acclimatize.

Mechanical Ventilation and Heat Recovery (MVHR)

As discussed previously, the air-tightness of the building envelope is a key consideration of any low energy home. As a building becomes air-tight the management of ventilation becomes more important. Stale moist air needs to be replaced by dry clean air to control the internal environment of the building. A build up of moisture inside a house can damage the fabric of the building and

encourage the growth of mould. Stale air can contain high levels of pollutants and CO_2, which have been proven to be damaging to health. In the past this was not so much of a problem as buildings were generally quite draughty but in a modern air-tight house a system to ventilate the building needs to be included at the design stage. Currently buildings tend to be ventilated by the use of bathroom and kitchen extractor fans and the clean air is supplied by trickle vents in the windows. In the summer this solution can be effective but in the winter energy is lost by cold air entering the house. The modern solution to this is to use a mechanically ventilated heat recovery (MVHR) system.

An MVHR system draws in fresh cold clean air from the outside of the building and draws out warm moist stale air from the inside. Both sets of air pass through a heat exchanger, so the cold outside air is pre-heated by the exiting warm internal air. The air is circulated through the building by a series of ducts. The exhaust air is removed from the kitchen and bathrooms and the intake air is distributed to the living spaces and bedrooms. Certain areas of the house such as stairwells and corridors have no air extract or supply and act as transfer paths for the movement of air through the building. The amount of air coming into the building has to be equal to the amount removed to make the system run efficiently. The air is delivered or removed via terminals set in the ceiling of the rooms where they are required.

Top left **Floorboards need to be at the right moisture content before they are laid out above the underfloor-heating pipes**

Top right **An example of an MVHR system**

Above **MVHR ducting running through open web joists**

Far right **A sandblaster in action**

Right **The lower half of this frame has been sandblasted whilst the upper section still waits to be cleaned**

The air is delivered at a steady low pressure rate so that drafts and noise from the terminal cannot be detected. The MVHR unit is also fitted with filters so that the supply air has pollens and pollutants removed before entering the house. This improves the air quality and can be a major advantage to people who suffer from asthma or hay fever. MVHR systems can be remarkably efficient and can supply air to a room at a temperature of 16.5 degrees when the air temperature outside is -10 degrees. But on the coldest days the supply air could need to be pre-heated. This is done by fitting a small heater to the supply air duct just after it leaves the MVHR unit. Good MVHR systems run at very low energy costs.

The design of the MVHR system should be done by a professional and incorporated in the construction drawings at an early stage. This is especially important in an oak-framed building where it might be difficult to conceal pipe runs. A service void needs to be created which bypasses the frame. This is normally done by using open web joists placed on top of the floor beams in the frame (see diagram). The size of the joists are determined not only by the span they have to achieve but also by the diameter of the ducting they have to conceal. Duct runs need to be as short and as straight as possible to improve the efficiency of the system and reduce any noise. The position of the MVHR unit should be placed as centrally as possible to aid this. Vertical duct runs need special consideration in an oak frame as it is not possible to pass them through a beam at any point. They are normally concealed in the divisions between rooms which need to be made wider than usual to accommodate the ducting.

Cleaning Green Oak

When oak is delivered from a sawmill, it invariably has blue-black stains on it. These are produced by the iron in the saw blades reacting with the natural tannin in the oak. In the workshop further marks are put on the oak by the carpenters snapping chalk lines and facing the timber. The finished beams are then transported to site which is often muddy and wet, causing them to get even dirtier. If it rains after the frame has been erected, brown water stains are produced on it by water leeching out the tannin in the oak. Add to all of this the dirty marks left by builders fixing the external envelope, and it's no wonder the frame needs cleaning.

Sandblasting

The most common way to clean the oak frame is by sandblasting. Doing it by other methods, such as manual sanding or planing each face, takes an inordinately long time and can prove to be costly. Sandblasting isn't cheap, though, so the cost should be factored into the build at the planning stage. Quotes can usually be supplied by specialist sandblasting companies, working from the framing drawings. Sandblasting works by shooting grit, fed from a hopper, at an extremely high pressure (100psi or 7.3kgcm^2) through a nozzle. When the grit hits the surface of the oak, it removes a thin layer of wood, leaving clean, bare timber behind. This can appear quite white and a bit patchy just after the sandblasting has been completed, but within a month the surface of the oak re-oxidizes and begins to turn an even golden colour. The blasting needs to be done with quite a light touch (if such a thing is possible) so the grain of the oak is just slightly raised but not blasted so much that it resembles driftwood. This is achieved by jetting the grit in a sweeping motion along the direction of the grain, never letting the nozzle rest in one place.

The type of grit used for blasting green oak is important, and disastrous results can occur if the wrong product is used. As previously mentioned, iron will react with tannin to create a black stain, and unfortunately most types of sandblasting grit contain small iron particles. If an oak frame is blasted with a grit containing iron particles, initially everything will appear to be fine, but after a couple of days the whole frame will start to turn black, as iron particles blasted into the frame react with the tannin. Once this has happened it can be difficult to reverse, as some of the particles can be deeply embedded in the oak and re-blasting will not always remove them. Luckily there is a non-ferrous grit available called 'SC', which is clean and safe to use. If there is any doubt about the grit, a sample should be left on some green oak for a week to check for any reaction. After this time, if there are any signs of discolouring it should not be used. Given the possible consequences of using the wrong type of grit, only sandblasters with a proven record and thorough understanding of blasting green-oak frames should be used.

The timing of the sandblasting is quite critical within the overall building process. It is, after all, the finishing treatment for the frame after which no further beautification is needed. The problem is, though, that grit blasted out at a great speed can easily cause collateral damage to materials placed

next to the oak. This is especially a problem with soft materials such as plasterboard, which may have been placed on top of the rafters prior to roofing. Any 'over-blast' when doing the rafters could easily rip a hole through it. On the other hand, if the frame is left unprotected after it's blasted it could be covered in water stains from rain in no time. One solution is to cover the frame with heavy-gauge polythene, which can easily be acquired from a builders' merchant, soon after it is erected. Once the frame has had a chance to dry out for a week or two, it can be blasted without fear of ruining any other materials, and if a hole is put in the polythene, it can be repaired simply with some gaffer tape. The polythene can then be removed section by section as the external envelope is built. If the rafters are covered with a timber sarking it is possible to blast the frame after the roof is watertight. There will inevitably be damage to the sarking, if it has been pre-painted, but it is possible to touch this up after the blasting has finished. Sandblasting, by its nature, is a messy, noisy business and requires a fair amount of clearing up afterwards. With all this in mind, make contact with a suitable sandblaster well in advance to co-ordinate the best time to blast the frame.

Steam, Oxalic Acid and Sanding

Whichever method is used to protect the frame after sandblasting, it is almost impossible not to get a few water stains in the oak before the envelope is completely watertight. These can be washed out with a scrubbing brush and some warm, soapy water. The whole beam will need to be done and not just the stain, otherwise a 'tide mark' will be left. A better way to clean them off is by using a steam cleaner. Hardware stores often sell wallpaper strippers that come with various attachments, one of which is used for cleaning upholstery. This can be wrapped in a cloth and used to clean off the water stain by rubbing it over the affected area. Because the steam evaporates immediately, no subsequent stain or tide mark is left. Steam is very hot, so protective clothing should be worn.

Oxalic acid is a chemical used in the dry-cleaning industry and is found naturally in rhubarb leaves. It is sold in the form of a crystalline powder from drugstores and chemical suppliers. It is very effective at removing tannin and water stains. The crystals are diluted with warm water, and painted onto the affected beam. Once again it is better to cover the whole beam, otherwise a tide mark will be left. After it has dried the beam should be washed with clean water to neutralize any residue. Oxalic acid is very caustic, so full protective clothing should be worn

when handling it and the manufacturer's instructions followed. It is best to use it over a small area only and test it first on a hidden section.

A lot of stains and marks can be removed with just a simple piece of sandpaper and some elbow grease. It is best to use sandpaper with a coarse grit and rub in the direction of the grain. Larger areas can be cleaned using a belt sander or a disc sander attached to an angle grinder. If the frame has been previously sandblasted, it can be difficult to remove water-stain marks without sanding away a lot of timber. This can change the appearance of the surface so a small area should be tested first. Goggles and a dust mask should always be worn when sanding oak.

Below **Water stains on the oak after the frame has been sandblasted, due to roof leakage. Stains like these can be cleaned using steam**

Far right **A well-finished oak-framed building will exist harmoniously with its surroundings, not just in terms of its visual impact but also its environmental impact**

Bottom **Barn-style oak frame with a direct glazed gable**

Finishing

∧ **Glazing frames** – Glazing should create a balance between the heat lost and the energy saved from using daylight as opposed to electric light. This is approximately 20 per cent and 30 per cent of the total floor area. Passive solar heating improves the thermal efficiency of the building by retaining the heat which is naturally produced from sunlight.

∧ **Floors** – A wide range of flooring systems can be used and each should include thermal insulation, acoustic insulation, a service zone, underfloor heating and a correct thermal mass.

∧ **Underfloor heating** – This has major benefits for oak-framed houses as it not only works at a much lower temperature, but also creates a more even temperature distribution than traditional radiators.

∧ **Sandblasting** – This is the common way to clean an oak frame and is a relatively speedy process. It isn't without its problems, however, and you should use a sandblaster who is experienced in meeting the challenges posed by oak-framed buildings.

∧ **Steaming, oxalic acid and sanding** – A few water stains will always creep into the frame after it has been sandblasted, and you will need to remove these by steaming, using oxalic acid or sanding.

∧ **MVHR** – Controlling ventilation in an air-tight building is very important but space for ducting needs to be incorporated into the design at an early stage.

Further Reading

Advanced Timber Framing
Steve Chappell
Fox Maple Press, 2013
ISBN: 978-1-88926-903-0

Build a Classic Timber-Framed House
Jack A Sobon
Storey Publishing, 1994
ISBN: 978-0-88266-841-3

Building Your Own Home
David Snell and Murray Armor
Random House UK, 18th edition 2006
ISBN: 978-0-09188-619-6

Conservation of Timber Buildings
F W B Charles and Mary Charles
Routledge, 2nd edition 1995
ISBN: 978-1-87339-417-5

The Conversion & Seasoning of Wood
William H Brown
Stobart Davies Ltd, 1987
ISBN: 978-0-85442-037-7

Discovering Timber-Framed Buildings
Richard Harris
Shire Publications Ltd, 3rd revised edition 1993
ISBN: 978-0-74780-215-0

Green Woodwork: Working with Wood the Natural Way
Mike Abbott
Guild of Master Craftsman Publications Ltd, 1998
ISBN: 978-0-94681-918-8

Hardwoods in Construction
C J Mettem & A D Richens
TRADA Technology Ltd, 1991
ISBN: 978-0-90134-883-8

The Housebuilder's Bible
Mark Brinkley
Ovolo Books Ltd, 10th edition 2013
ISBN: 978-1-90595-915-0

How Structures Work: Design and Behaviour from Bridges to Buildings
David Yeomans
Wiley-Blackwell, 2009
ISBN 978-1-40519-017-6

Oak: A British History
Esmond Harris, Jeanette Harris and N D G James
Windgather Press, 2003
ISBN: 978-0-95386-308-2

The Passivhaus Handbook
Janet Cotterell and Adam Dadeby
Green Books, 2012
ISBN: 978-0-85784-019-6

Reciprocal Frame Architecture
Olga Popovic Larsen
Architectural Press, 2008
ISBN 978-0-7506-8263-3

Shelter
Lloyd Kahn (ed)
Shelter Publications Inc, 2nd edition 2000
ISBN: 978-0-93607-011-7
www.shelterpub.com

Simplified Design of Wood Structures
James Ambrose, Patrick Tripeny
John Wiley & Sons Inc., 6th edition 2009
ISBN: 978-0-47018-784-5

The Repair of Historic Timber Structures
David J Yeomans
Thomas Telford Ltd UK, 2003
ISBN: 978-0-72773-213-2

Timber Building in Britain
R W Brunskill
Yale University Press, new edition 2006
ISBN: 978-0-30436-665-1

Timber Framing
Journal of the Timber Framers' Guild
www.tfguild.org

The Timber-Frame Home
Tedd Benson
Taunton Press USA, 2nd edition 1997
ISBN: 978-1-56158-129-0

A Timber Framer's Workshop
Steve Chappell
Fox Maple Press Inc. USA, 3rd revised edition 2007
ISBN: 978-1-88926-900-9

Your Home Technical Manual
Australia's Guide to Environmentally Sustainable Homes
Chris Reardon et al
Australian Government Department of Climate Change and Energy Efficiency, 4th edition 2010
ISBN: 978-1-921299-20-9

Useful Contacts

Association for Environment Conscious Building
(AECB)
www.aecb.net

Forest of Avon Products
www.forestofavonproducts.co.uk

Forestry Commission, UK
www.forestry.gov.uk

Forestry Stewardship Council (FSC)
www.fsc.org

Guidance on planning permission in the UK
www.plannningportal.gov.uk

Health and Safety Executive
www.hse.gov.uk

Land Registry in the UK
www.landregistry.gov.uk

Land Registry in the USA
www.uslandrecords.com

National Hardwood Lumber Association
www.nhla.com

Occupational Safety & Health Administration
www.osha.gov

Timber Framers Guild
www.tfguild.org

The Timber Research and Development Agency
(TRADA)
www.trada.co.uk

The Carpenters Fellowship
www.carpentersfellowship.co.uk

Westwind Oak Buildings Ltd
www.westwindoak.com

Places to Visit

Avoncroft Museum of Historic Buildings, Worcs, UK
www.avoncroft.org.uk

Centre for Alternative Technology, Powys, UK
Also run courses in green-oak framing
www.cat.org.uk

Chiltern Open Air Museum, Bucks, UK
www.coam.org.uk

The Gateway Centre, Cotswold Water Park,
Gloucs, UK
www.waterpark.org

King John's Hunting Lodge, Axbridge, Somerset, UK
www.nationaltrust.org.uk/king-johns-hunting-lodge

Llandoger Trow, Public House, Bristol
3–5 King Street, Bristol BS1 4ER, UK

St Fagans National History Museum, Cardiff, UK
www.museumwales.ac.uk/stfagans

Weald & Downland Open Air Museum,
West Sussex, UK
Also run courses in green-oak framing
www.wealddown.co.uk

Suppliers and Manufacturers of Tools

Axminster Power Tools
www.axminster.co.uk

Barr Tools, Idaho, USA
www.barrtools.com

Bristol Design (old tools)
www.bristol-design.co.uk

Cromwell Tools
www.cromwell.co.uk

Henry Taylor Tools Ltd
www.henrytaylortools.co.uk

NMA Agencies (Mafell) UK
www.nmatools.co.uk

Robert Sorby Tools
www.robert-sorby.co.uk

Tilgear
www.tilgear.info

Glossary

Aisled frame – A frame which has aisles added to a wall frame to increase the span of the building.

Arcade posts – Also known as aisle posts, are the internal posts in an aisled frame.

Arch-brace truss – Truss with curved braces which joint into a collar. A common truss found in UK churches.

Arris – The corner edge of a beam.

Bay – The space between cross frames.

Birdsmouth – A right-angled cutout on the underside of a rafter, where it notches over the wallplate. The depth of the birdsmouth should be a third of the depth of the rafter.

Box frame – A system of framing where posts and wall plates support roof trusses.

Brace – Diagonal piece timber to add strength and stop the building from racking. In the UK these are usually cut out of curved timber.

Bressumer – Sill supporting the upper wall above a jetty.

Bridging beam – Large floor beam supporting ends of joists.

Bridle scarf – Scarf joint employed in joining wallplates and soleplates.

Cambered beam – Centre of the beam is higher than the ends.

Carpenters' mark – A chiselled reference number put on each timber when it is made.

Chamfer – A surface made on the arris of a beam by taking off a 45° flat piece of timber.

Cleat – A small piece of wood attached to the back of a principal rafter to stop any rotation in the purlin.

Collar – A horizontal timber between the principal rafters in a truss.

Collar purlin – Alternative name for a crown plate.

Common rafter – Rafters which are supported by the purlins.

Cripple-jack rafter – A rafter that is connected to both a hip and valley rafter.

Crown plate – Central longitudinal beam in a roof that supports collared rafters.

Crown post – Post which supports the crown plate. They are usually shaped for decoration.

Cruck blades – Curved pairs of large timbers, which reach from the floor to the ridge. Usually a matching pair is made by splitting a curved tree in half.

Dovetail – A locking wedge-shaped joint.

Dragon beam – Floor beam set diagonally to support a jettied floor.

Draw-pegging – Method by which the peg hole in the tenon does not quite line up with the peg hole in the mortise, so when a tapered peg is driven into the joint, it closes up tightly.

Gable end frame – The outside end cross frame of a building.

Girding beam – Horizontal beam known as girth or rail, at level of upper floor.

Green oak – Freshly felled oak still with high moisture content.

Hip rafter – A diagonal roof timber that rises from an external corner of a frame, to the ridge.

Interrupted tie beam truss – A truss where a section of tie beam is cut and then jointed into a pair of queen posts. These then transfer the horizontal thrust from the interrupted tie beams into a lower floor beam.

Jack rafter – A rafter which is connected to the hip rafter.

Jettied building – A building in which each floor overhangs the floor below. Seen in many timber-framed town houses and also Wealdon-style buildings.

Joists – Horizontal beams in the floor that support the floorboards.

Jowl post – Flared head of main post, which enables the joining of wallplate and tie beam. Also known as root-stock because they are made from the bottom of the tree where the root bowl swells.

King post – Central post in a truss used to support the tie beam.

Lap joints – Halving joints which cross over each other.

Mortise and tenon – Common joint for connecting two pieces of timber together.

Peg – Wooden pin used to connect the joints in a frame. Also know as a tree-nail.

Primary timber – A timber that is common to more than one two-dimensional plane.

Principal rafter – The main diagonal beams in trusses that support the purlins in the roof.

Purlin – Horizontal roof beams which support the rafters and carry the roof load into the cross frames.

Queen posts – A pair of timbers in a truss, used to support the principal rafter by the purlins. These can be curved or straight.

Racking – An effect where one beam or post pushes its connecting member until the structure collapses like a stack of dominoes. Braces are used in frames to counteract the affects of racking.

Ridge – Horizontal timber supporting top ends of rafters in the apex of the roof.

Roof truss – 'A' frames in the roof that support the purlins.

Scarf joint – Used to connect two lengths of timber together to form one continuous length.

Scribing – A method of marking out joints on out of square timber.

Secondary timber – A timber that is common to only one two-dimensional plane.

Shake – A fissure or split in a beam.

Sill – Horizontal bottom timber of window frame.

Sling brace – A long curved brace used in an interrupted tie beam truss.

Sole plate – A continuous timber which runs around the bottom of the frame.

Spandrel – Space between jointed post, brace and beam.

Sprocket – A rafter extension which is usually fixed at a different angle to the rafter to 'kick' up the bottom row of tiles.

Strut – A timber that projects from a king post to support the principal rafters.

Studs – Also know as quarters, studs are small vertical posts within the frame.

Tie-beam – Large horizontal beam that ties wallplates together and also forms main part of a truss.

Transom – Small horizontal timbers jointed into the studs to form openings.

Trusses – 'A' frames in the roof which take the load from the purlins.

Valley rafter – A diagonal roof timber that rises from the internal corner of a frame.

Wallplate – A continuous timber that runs along the top of a wall frame.

Wind brace – Brace from principal rafter to purlin that strengthens the roof structure and stops the trusses from racking.

Yoke – A small timber attached to the apex of a truss to support the ridge.

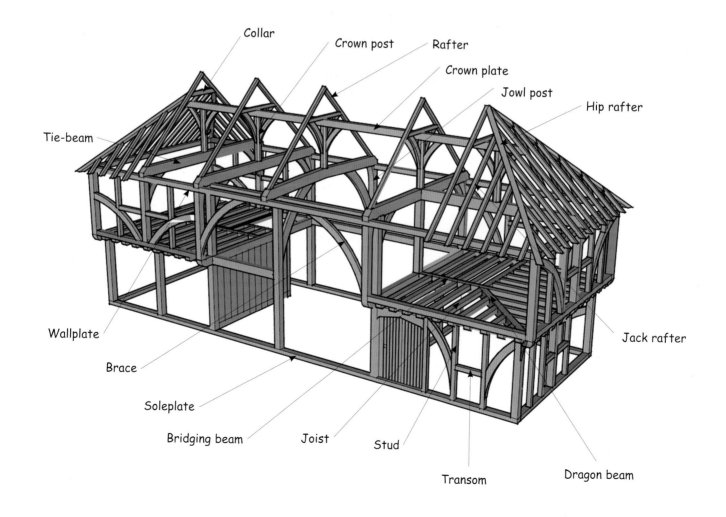

Index

To place an order, or to request a catalogue, contact: GMC Publications Ltd
Castle Place, High Street, Lewes, East Sussex BN7 1XU, UK
Tel: +44 (0) 1273 488005 www.gmcbooks.com